Jossey-Bass Teacher

Jossey-Bass Teacher provides educators with practical knowledge and tools to create a positive and lifelong impact on student learning. We offer classroom-tested and research-based teaching resources for a variety of grade levels and subject areas. Whether you are an aspiring, new, or veteran teacher, we want to help you make every teaching day your best.

From ready-to-use classroom activities to the latest teaching framework, our value-packed books provide insightful, practical, and comprehensive materials on the topics that matter most to K–12 teachers. We hope to become your trusted source for the best ideas from the most experienced and respected experts in the field.

D1377462

Math Wise!

Over 100 Hands-On Activities that Promote
Real Math Understanding, Grades K–8

Second Edition

Jim Overholt
Laurie Kincheloe

JOSSEY-BASS
A Wiley Imprint
www.josseybass.com

Published by Jossey-Bass
A Wiley Imprint
989 Market Street, San Francisco, CA 94103-1741—www.josseybass.com

Jossey-Bass books and products are available through most bookstores. To contact Jossey-Bass directly call our Customer Care Department within the U.S. at 800-956-7739, outside the U.S. at 317-572-3986, or fax 317-572-4002.

Jossey-Bass also publishes its books in a variety of electronic formats. Some content that appears in print may not be available in electronic books.

ISBN: 978-0-470-471999

Printed in the United States of America

SECOND EDITION
PB Printing   10 9 8 7 6 5 4 3

About This Resource

Math Wise! includes activities that will help each student gain full comprehension of basic mathematical concepts, including numbers and counting, computation, estimation, probability, data analysis, measurement, geometry, algebra, problem solving, and logical thinking. Students in today's math classrooms must be able to do more than achieve correct answers through computation; they need to understand basic concepts and experience a range of mathematical applications. *Math Wise!* is designed to help the teacher accomplish these learning objectives. It contains a wide variety of learning experiences that have been arranged according to difficulty level. Whenever possible, the activities are presented in either hands-on or visual formats.

Concrete/Manipulative Activities

Especially when exploring "new" concepts, each student should work with hands-on materials. A number of the activities therefore include easily obtained manipulatives, such as straws, paper clips, sugar cubes, and beans. For example, a problem in the activity *Paper Clip Division* asks students to show 44 divided by 7. One-to-one correspondence is used when one paper clip corresponds to the numeral 1. The result might appear as:

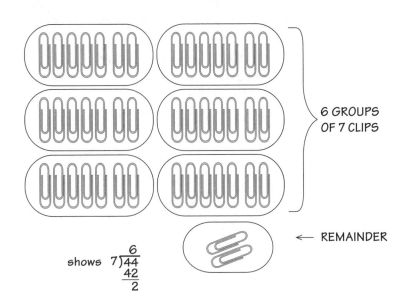

6 GROUPS OF 7 CLIPS

← REMAINDER

shows $7\overline{)44}$ ... $\frac{6}{}$ ... $\frac{42}{2}$

In *Punchy Math,* students use a paper hole punch, scrap paper, and a pencil to show $3 \times 7 =$ ____. The outcome, after folding, punching, looping, and labeling, shows 3 groups of 7. If turned sideways, it can also show 7 groups of 3, or $7 \times 3 =$ ____. Whereas the resulting punched holes are concrete, the looped segments provide a visual component that directly corresponds to the abstract number relationships involved.

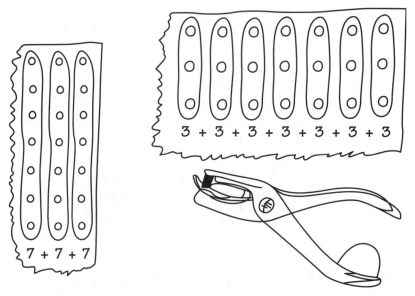

Such manipulative activities provide a basis for true understanding of mathematical concepts. For this reason, each section contains a number of similar exercises.

Visual/Pictorial Activities

For many learners, visual representations of mathematical problems are keys to the comprehension of these problems. Often visual representations involve 1-to-1 correspondence in connecting pictures with numbers. For example, in *Cross-Line Multiplication,* three horizontal lines represent the number 3 and five vertical lines represent the number 5. When the lines are crossed, the fifteen intersection points represent the answer to the problem $3 \times 5 =$ ____. The following figure illustrates this visual representation. Of course, turning the drawing sideways shows $5 \times 3 = 15$.

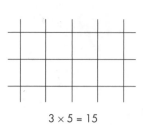

$3 \times 5 = 15$ $5 \times 3 = 15$

In *Decimal Squares,* another visual activity, students are provided with a sheet of Decimal Squares. Each decimal square is a 10-unit by 10-unit square divided into 100 square units. Each small square unit represents one hundredth of the decimal square, or .01. Students are then asked to show the relationship between 0.6 and 0.21. For example, in the problem 0.6 ____ 0.21, students are required to fill in the blank with >, <, or = to make the statement true. To find the answer, students are asked to shade in the Decimal Squares, as shown below.

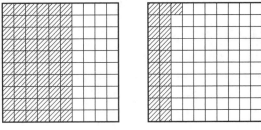

0.6 > 0.21

Abstract Procedures

A major goal of mathematics education is to help students eventually perform abstract mathematical procedures and understand the underlying concepts behind these procedures. When possible, mathematics teachers should not only instruct students in regard to mathematical mechanics but also enable them to gain a true understanding of the concepts involved.

In the activity *Post-it Mental Math,* one student has Post-it numerals placed on his or her back without being allowed to see them. The other group members, after viewing the numerals, give the student clues about the numerals. Using these clues, the Post-it wearer must do mental math to determine the numerals. In the situation that follows, the Post-it player has made a first guess based on one player's clues.

Block Four, which requires two players or two opposing teams, a numbered game board, and two paper clips, is another activity asking students to make abstract computations and draw upon their logical-thinking abilities. The first player places the paper clips on two numbers, and then performs the multiplication. The student then puts an X on the square with the answer. The next player can only move one paper clip, leaving the other one alone. This player will then perform the multiplication and mark his or her square with an O. The boards below show two partially played games.

BLOCK FOUR
Multiplication Facts

1	2	3	4	5	6
7	⑧	9	10	12	14
15	⑯	18	⊠	21	24
25	27	⊠	30	32	35
36	40	42	45	48	49
54	56	63	64	72	81

1 ② 3 ④ 5 6 7 8 9

BLOCK FOUR
Multiplication of Fractions

$\frac{1}{9}$	$\frac{2}{9}$	$\frac{21}{64}$	$\frac{1}{36}$	$\frac{1}{4}$	$\frac{5}{32}$	$\frac{5}{72}$	$\frac{1}{54}$
$\frac{1}{8}$	$\frac{25}{48}$	$\frac{1}{32}$	$\frac{3}{16}$	$\frac{1}{36}$	$\frac{7}{24}$	$\frac{1}{6}$	$\frac{7}{72}$
$\frac{1}{12}$	$\frac{1}{2}$	$\frac{3}{32}$	$\frac{1}{9}$	$\frac{5}{24}$	$\frac{1}{16}$	$\frac{7}{32}$	$\frac{1}{16}$
$\frac{5}{16}$	$\frac{7}{48}$	$\frac{1}{24}$	$\frac{5}{48}$	$\frac{3}{32}$	$\frac{15}{64}$	$\frac{1}{6}$	$\frac{4}{9}$
$\frac{9}{64}$	$\frac{1}{8}$	$\frac{2}{27}$	$\frac{7}{12}$	$\frac{1}{3}$	$\frac{1}{27}$	$\frac{1}{24}$	$\frac{7}{16}$
$\frac{15}{32}$	$\frac{5}{12}$	$\frac{9}{32}$	$\frac{7}{64}$	$\frac{1}{12}$	$\frac{21}{31}$	$\frac{1}{4}$	$\frac{1}{64}$
$\frac{1}{18}$	$\frac{9}{16}$	$\frac{25}{64}$	$\frac{3}{8}$	$\frac{1}{48}$	$\frac{49}{64}$	$\frac{3}{64}$	$\frac{3}{16}$
$\frac{1}{72}$	$\frac{5}{24}$	$\frac{1}{18}$	$\frac{5}{64}$	$\frac{1}{8}$	$\frac{1}{18}$	$\frac{35}{64}$	$\frac{1}{12}$

$\frac{1}{2}$ $\frac{1}{3}$ $\frac{2}{3}$ $\frac{1}{4}$ $\frac{3}{4}$ $\frac{1}{8}$ $\frac{3}{8}$ $\frac{5}{8}$ $\frac{7}{8}$ $\frac{1}{6}$ $\frac{1}{9}$

A Final Note

Students will find the activities and investigations from this book informative, interesting, and fun. Most important, students will gain a better understanding of the mathematics they are expected to master. *Math Wise!* will prove to be a most valuable supplement to any mathematics program.

Jim Overholt
Laurie Kincheloe

About the Authors

James L. Overholt has an Ed.D. from the University of Wyoming, Laramie. He has been exploring the use of manipulative and visual materials for mathematics instruction since the 1960s. As an elementary and secondary school teacher in Minnesota and Wyoming, and later as a university professor, his investigations have taken him into both K–12 classrooms and adult mathematics learning workshops. He is currently a professor of education at California State University, Chico.

Dr. Overholt regularly conducts mathematics education courses and workshops for pre-service and in-service teachers at the elementary and secondary levels. His earlier published books include *Math Stories for Problem Solving Success, Second Edition,* also published by Jossey-Bass/Wiley; *Dr. Jim's Elementary Math Prescriptions; Math Problem Solving for Grades 4–8; Math Problem Solving for Beginners Through Grade 3; Outdoor Action Games for Elementary Children,* and *Indoor Action Games for Elementary Children.*

Laurie Kincheloe has a B.A. in mathematics and an M.A. in mathematics education from California State University, Chico. She taught high school mathematics for twelve years and is presently teaching mathematics at Butte College in northern California. She has worked with K–12 students, parents, and teachers as a family math coordinator and as a mentor for new teachers. She teaches concepts in mathematics to pre-service elementary teachers, and has coordinated service learning projects connecting high school and college students with elementary students through mathematics. She was co-coordinator of the Mathematics Project at California State University, Chico, and has conducted workshops on the teaching of mathematics for elementary and secondary teachers at numerous education conferences.

In addition to teaching at Butte College, Laurie has served as the developmental coordinator for the Mathematics Department, created a math-anxiety class designed to help apprehensive students be successful at math, and organized the annual Math Awareness Week. She has received the Faculty Member of the Year Award and the Service Learning Project Faculty Award.

A Special Acknowledgement:
James F. Lindsey, Ed.D. (University of California, Berkeley) served as
an elementary teacher and principal for 25 years. He co-authored
Math Stories for Problem Solving Success: Ready-to-Use Activities
for Grades 6–12, First and Second Editions *(Jossey-Bass/Wiley).*
When asked if he would help edit and proofread the new edition of
Math Wise!, *he remarked "I would be honored!" From beginning to*
end, James was always ahead of expectations. He will be missed.

Suggestions for Using Math Wise!

The activities in this book provide a varied collection of interesting and understandable tasks from which students in kindergarten through the middle grades will benefit. Although many of these activities can be used in any order, it is advisable to designate tasks that are appropriate with regard to class size, students' stages of learning, or other considerations. For this reason, several features in this book are designed to help select appropriate activities.

- The **Contents** categorizes each activity in five ways:
  1. *Section* ("Making Sense of Numbers," "Computation Connections," "Investigations and Problem Solving," and "Logical Thinking")
  2. *Descriptive Title* (such as *Everyday Things Numberbooks*, *Paper Clip Division*, *Peek Box Probability*, and *String Triangle Geometry*)
  3. *Grade Level* (K–2, 2–4, 4–6, and 6–8)
  4. *Activity Type* (Concrete/Manipulative, Visual/Pictorial, and Abstract)
  5. *Learning Format* (Total Group, Cooperative, and Independent)

- A **Key** for each activity notes the most appropriate grade levels, the preferred working arrangement, and the kinds of experiences in which learners will take part. For example, the following key to *Silent Math* indicates that
  1. The activity is best suited for students in grades 4 through 8.
  2. The activity can be worked on by the whole class or by cooperative groups.
  3. The students will work with visual diagrams and will perform abstract computations.

Silent Math

Grades 4–8

☒ Total group activity
☒ Cooperative activity
☐ Independent activity
☐ Concrete/manipulative activity
☒ Visual/pictorial activity
☒ Abstract procedure

- Each activity begins with a **Why Do It** statement that details the specific mathematical concepts the students will be learning and practicing.

- The **You Will Need** statement specifies any supplies or equipment necessary for the activity. These items, such as paper clips, index cards, and straws, are easily obtained and free or inexpensive.

- The **How To Do It** section details what the teacher or other education professional must do to set up and carry out the activity. Suggestions are made as to the steps that should be taken for the activity to be successful. It also describes how the investigation works best as an independent activity, a cooperative project where students work in pairs or small groups, or a total group venture. This section will provide the general premise and content of the activity before the example are presented.

- The **Examples** illustrate how the activity might progress, and display typical outcomes.

- An **Extensions** section at the end of each activity contains more investigations that can be done using the same or similar procedures described in the activity. It often contains more sample questions or suggestions as to how to expand the mathematical concepts being studied. Teachers and students are encouraged to propose similar tasks of their own.

- Where appropriate, reproducible pages immediately follow the relevant activity. These pages include game boards, workmats, dot paper, playing cards, graph paper, and more.

- Students should be encouraged to record their methods and solutions in a math journal or to keep a special file containing samples of their work.

- Solutions are also provided when appropriate.

Contents

	Grade Level				Activity Type				Learning Format			
	K–2	2–4	4–6	6–8		Concrete/Manipulative	Visual/Pictorial	Abstract		Total Group	Cooperative	Independent	
50. Number Grids		X	X	X				X		X	X	X	186
51. Here I Am		X	X	X				X		X	X		189
52. Equation Match-Up		X	X	X				X		X	X		194
53. Block Four		X	X	X				X			X		196
54. Silent Math			X	X			X	X		X	X		203
55. Rapid Checking			X	X				X		X	X	X	206
III. Investigations and Problem Solving													209
56. Shoe Graphs	X	X				X	X	X		X	X		211
57. Sticky Gooey Cereal Probability	X	X				X	X	X			X	X	214
58. Sugar Cube Buildings	X	X	X	X		X	X			X	X	X	219
59. A Chocolate Chip Hunt	X	X	X	X		X	X	X		X	X	X	223
60. Flexagon Creations	X	X	X	X		X	X			X	X	X	228
61. Watermelon Math	X	X	X	X		X	X	X		X	X		232
62. Restaurant Menu Math	X	X	X	X			X	X		X	X	X	235
63. Peek Box Probability	X	X	X	X		X	X	X		X	X	X	238
64. A Problem-Solving Plan	X	X	X	X		X	X	X		X	X	X	242
65. Fraction Quilt Designs	X	X	X	X		X	X	X		X	X	X	247
66. What I Do in a Day		X	X				X	X		X		X	250
67. Shaping Up		X	X	X		X	X	X		X	X	X	254
68. Verbal Problems		X	X	X				X		X	X	X	260
69. Scheduling		X	X	X			X	X		X	X	X	271
70. Student-Devised Word Problems		X	X	X				X		X	X	X	274
71. Tired Hands		X	X	X		X	X	X		X	X		278
72. Paper Airplane Mathematics		X	X	X		X	X	X		X	X	X	281
73. A Dog Pen Problem		X	X	X		X	X	X		X	X	X	285
74. Building the Largest Container		X	X	X		X	X	X		X	X	X	288
75. The Three M's (Mean, Median, and Mode)		X	X	X			X	X		X		X	290
76. Post-it Statistics		X	X	X		X	X	X		X	X		294
77. A Postal Problem			X	X		X	X	X		X	X	X	297

Contents

Contents

Section One

Making Sense of Numbers

The activities in this section introduce students to many number concepts and relationships, including 1-to-1 correspondence, basic number combinations, place value, mental math, fractions, large numbers, and decimals. Students will practice essential mathematical skills and develop conceptual understanding through hands-on investigations and games that make use of manipulative experiences, visual portrayals, or relevant abstract procedures.

A number of activities from other portions of this book can be used to extend and enhance students' comprehension of the concepts introduced in this section, such as *Punchy Math* (p. 104) and *Beat the Calculator* (p. 122) from Section Two; *Peek Box Probability* (p. 238) and *Restaurant Menu Math* (p. 235) from Section Three; and *Duplicate Digit Logic* (p. 408) from Section Four.

Chapter 1

Toothpick Storybooks

Grades K–3
☒ Total group activity
☒ Cooperative activity
☒ Independent activity
☒ Concrete/manipulative activity
☒ Visual/pictorial activity
☒ Abstract procedure

Why Do It:

Students will discover the concepts of 1-to-1 counting and number conservation, and will study basic computation relationships.

You Will Need:

This activity requires several boxes of flat toothpicks, white and colored paper (pages approximately 6 by 9 inches work well), glue, and marking pens or crayons.

How To Do It:

1. Have younger students explore and share the different arrangements they can make with a given number of toothpicks. For example, students could arrange 4 toothpicks in a wide variety of different configurations, all of which would still yield 4 toothpicks.

2. After exploring for a while, students should begin making Toothpick Storybooks, starting by creating number pages. Students can write, for instance, the number 6 on a sheet of white paper and glue 6 toothpicks onto a piece of colored paper. (To avoid a sticky mess, students should dip only the ends of the toothpicks in the glue.)

When they are ready, the learners follow the same procedure for equations and the corresponding toothpick pictures. (*Note:* Students sometimes portray subtraction by pasting a small flap on the colored page that covers the number of toothpicks to be "taken away." Furthermore, they enjoy lifting the flap to rediscover the missing portion.)

3. When a number of toothpick diagrams have been finished, the pages can be stapled together into either individual or group Toothpick Storybooks. Ask each student to tell a number story about one of the diagrams in which he or she makes reference to both the toothpick figure and the written equation or number.

Example:

Shown here are possible toothpick diagrams for 4, $3 + 5 =$ ____, and $7 - 2 =$ ____.

Extensions:

1. Simple multiplication facts, and even longer problems, can be portrayed with toothpick diagrams. For $6 \times 3 =$ ____, the player might show ||| ||| ||| ||| ||| ||| = 18. Similarly, for $4 \times 23 =$ ____, it is necessary to show 4 groups of 23 toothpicks to yield 92.

2. Division can also be shown with toothpick diagrams. If the problem calls for the division of 110 into sets of 12, the player would need to form as many groups of 12 as possible, also taking into account any remainder. (*Note:* The student might also complete such a problem using partitive division. See *Paper Clip Division*, p. 179.)

Chapter 2

Number Combination Noisy Boxes

Grades K–3

☐ Total group activity
☒ Cooperative activity
☒ Independent activity
☒ Concrete/manipulative activity
☒ Visual/pictorial activity
☒ Abstract procedure

Why Do It:

This activity provides students with a visual and concrete aid that will help them understand basic number combinations and practice addition and subtraction.

You Will Need:

Ten (or more) stationery or greeting card boxes with clear plastic lids, approximately 50 marbles, and pieces of Styrofoam or sponge that can be trimmed to fit inside the boxes are required.

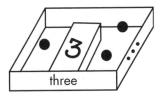

How To Do It:

1. Construct Noisy Boxes for the numerals 0 through 9 (or beyond). For each box, cut the foam to make a divider that will lie perpendicular to the bottom of the box. Glue the divider to the bottom of the box, ensuring that it is trimmed down such that the marbles

will pass over it when the top is on (see figure). Use a marking pen to write the numeral, such as 3, on the divider and to inscribe the appropriate number of dots on one outside edge of the box (● ● ●) and the corresponding number word on another outside edge (three). Insert that same number of marbles into the box and tape on the clear plastic lid.

2. Allow the students to work with different Noisy Boxes. Instruct students to tip or shake a Noisy Box so that some or all of the marbles roll past the divider. Once this is done, the player is to record the outcome as an addition problem. The student should shake the same Noisy Box again and record a new outcome. For example, three marbles will yield outcomes such as $1 + 2$, $3 + 0$, $2 + 1$, or $0 + 3$. The activity continues in this manner until no further combinations are possible (see Example below).

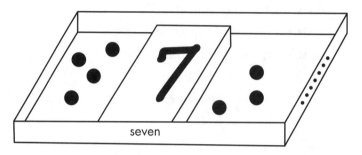

Example:

The recorded number combinations for the 7s Noisy Box should include the following:

Addition		Subtraction	
$4 + 3 = 7$	$6 + 1 = 7$	$7 - 4 = 3$	$7 - 6 = 1$
$3 + 4 = 7$	$1 + 6 = 7$	$7 - 3 = 4$	$7 - 1 = 6$
$5 + 2 = 7$	$0 + 7 = 7$	$7 - 5 = 2$	$7 - 0 = 7$
$2 + 5 = 7$	$7 + 0 = 7$	$7 - 2 = 5$	$7 - 7 = 0$

Extension:

If any player has difficulty on a visual level in utilizing a Noisy Box, have that student temporarily remove the plastic lid. Then he or she can touch and physically move the marbles from one side of the box to the other. Nearly all students will experience success as a result of such a tangible experience with number combinations.

Chapter 3

Everyday Things Numberbooks

Grades K–4
- ☒ Total group activity
- ☒ Cooperative activity
- ☒ Independent activity
- ☒ Concrete/manipulative activity
- ☒ Visual/pictorial activity
- ☒ Abstract procedure

Why Do It:

Students will discover that in their daily lives there are many things that come in numbered amounts, such as wheels on a bicycle.

You Will Need:

Each student will require paper that can be stapled into booklets, pencils, scissors, and glue sticks or paste (if desired).

How To Do It:

1. As a group, discuss things in everyday life that are generally found as singles or 1s: 1 nose for each person, 1 trunk per tree, 1 beak on a bird, 1 tail per cat, 1-a-day multiple vitamins, and so on. Then provide each student with a sheet of paper and have everyone write the number 2 at the top. Each participant should list as many things that come in 2s as he or she can think of, such as 2 eyes, ears, hands, and legs for each person; 2 wings per bird, and so on. Do the same for 3s: 3 wheels on a tricycle; 3 sides for

any triangle; a 3-leaf clover, and so on. Students might also paste pictures representing numbered amounts on their pages. Have them complete a page (or more) for each number up to 10 or larger, and then discuss their ideas. You may want to construct large class lists for each number. This activity can continue for several days, and may be assigned as homework.

2. At first some numbers seem unusable, but wait and you will be delighted with students' suggestions. For instance, 7 can be illustrated by 7-UP®, and 8 depicted by 8 sides on a stop sign. Students will often continue to make suggestions, even after the activity has ended!

Example:

The following is a partial Numberbook listing for the number 4.

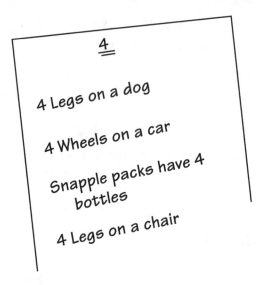

Extensions:

Ask more advanced students to consider the following problems:

1. What items can commonly be found in 25s, 50s, 100s, or any other number you or students might come up with? Is there any number for which an example cannot be found?

2. Find examples for fractional numbers. If there are 12 sections in an orange, 1 of those sections is 1/12 of the orange; 3 of those sections are 3/12 or 1/4 of the orange.

Chapter 4

Under the Bowl

Grades K–3

- ☐ Total group activity
- ☒ Cooperative activity
- ☒ Independent activity
- ☒ Concrete/manipulative activity
- ☒ Visual/pictorial activity
- ☒ Abstract procedure

Why Do It:

Under the Bowl provides students with a visual and concrete aid that will help them understand basic number combinations and practice addition and subtraction.

You Will Need:

A bowl or small box lid and small objects (such as beans, blocks, or bread tags) are required for each player.

How To Do It:

Allow students a brief period to explore their bowls and objects. Have students begin the activity with small numbers of items: students with 3 beans, for example, might be told to put 2 beans under the bowl and place the other on top of it. Then they should say aloud to a partner or together as a class, "One bean on top and two beans underneath make three beans altogether." Once students understand the activity, ask them to keep a written record of their work; for 3 beans, as noted above, they should record $1 + 2 = 3$ (after they have had instruction on four fact families, they should also record $2 + 1 = 3$, $3 - 2 = 1$, and $3 - 1 = 2$). Although initially students should use only a few objects, they might go on to use as many as 20, 30, or even 100 items.

Example:

The players shown above are working with 7 beans. Thus far they have recorded the four fact family for 1 bean on top of the bowl and 6 beans under it. They are now beginning to record their findings for 2 beans on top. Next they might put 3 beans on top and record. (*Note:* Should a student become confused about a number combination, he or she may count the objects on top and then lift the bowl to either visually or physically count the objects underneath. This usually helps clarify the problem.)

Extensions:

1. When older students are working as partners, an interesting variation has one student making a combination and the other trying to figure out what it is. For example, the first student might put 3 beans on top of the bowl and some others under it. He or she then states, "I have 11 beans altogether. How many beans are under the bowl, and what equations can you write to represent this problem?" The second student should respond that there are 8 beans under the bowl, yielding the equations $3 + 8 = 11, 8 + 3 = 11, 11 - 8 = 3$, and $11 - 3 = 8$.

2. You can also extend this activity to introduce algebra concepts to students. For example, after instruction a student presents the problem shown in Extension 1, with the equation $n + 3 = 11$. Explain to students that using a letter to represent a missing number is a basic concept in algebra.

Chapter 5

Cheerios™ and Fruit Loops™ Place Value

Grades K–5

☒ Total group activity

☒ Cooperative activity

☒ Independent activity

☒ Concrete/manipulative activity

☒ Visual/pictorial activity

☒ Abstract procedure

Why Do It:

Students will begin to understand place value concepts through a visual and concrete experience.

You Will Need:

This activity requires several boxes of Cheerios and one box of Fruit Loops breakfast cereals, string or strong thread, needles, and two paper clips (to temporarily hold the cereal in place) for each group or student. (*Note:* If you do not wish to use needles, you can use waxed or other stiff cord.)

How To Do It:

After a place value discussion about 1s, 10s, and 100s, challenge the students to make their own place value necklaces (or other decorations). Ask them to determine the "place value" of their own necks and, when they look puzzled, ask, "How many Cheerios on a string will it take to go all the way around your neck?" Then explain that they will be stringing Cheerios and Fruit Loops on their necklaces in a way that shows place value: for each 10 pieces of cereal to be strung, the first 9 will be Cheerios and every 10th piece must be a colored Fruit Loop. They will then be able to count the place value on their necks as 10, 20, 30, and so on. As students finish their necklaces, be certain to have students share the numbers their necklaces represent and how the necklace displays the number of 10s and the number of 1s in their number.

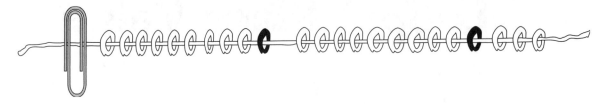

Example:

The partially completed Cheerios and Fruit Loops necklace shown above has the place value of two 10s and three 1s, or 23.

Extensions:

1. As a class, try making and discussing other personal place value decorations, such as wrist or ankle bracelets or belts.

2. An engaging group project is to have students estimate the length of their classroom (or even a hallway) and make very long Cheerios and Fruit Loops chains. Be sure to initiate place value discussions about 100s; 1,000s; and even 10,000s or more. (*Hint:* When making such long chains, it is helpful to have individuals make strings of 100 and then tie these 100s strings together.)

Chapter 6

Beans and Beansticks

Grades K–6

☒ Total group activity

☒ Cooperative activity

☒ Independent activity

☒ Concrete/manipulative activity

☒ Visual/pictorial activity

☒ Abstract procedure

Why Do It:

This activity will give students 1-to-1 concrete and visual understandings of place value and computation concepts.

You Will Need:

Required for this project are dried beans, Popsicle sticks (or tongue depressors), and clear-drying carpenter's glue (most forms of permanent glue work well).

How To Do It:

1. Beansticks, a stick with 10 beans, will help students visualize the 10s place of a number. A flat, a raft made of 10 beansticks, will allow students to view the 100s place of a number. Single beans will serve for the numbers 1 through 9, but 10s beansticks and 100s flats (or rafts) will be needed after that. The 10s beansticks are constructed by gluing ten beans to a stick, and the 100s flats are made by gluing ten beansticks together

with cross supports (see the illustrations below). The beansticks and rafts will be more durable if a second bead of glue is applied several hours after the first layer of glue dries. Finally, to enhance the lesson, students should construct the beansticks themselves; or, if necessary, have older students help younger students with the construction and then use the beansticks to lead a lesson in place value.

2. The students should first use single beans to represent single-digit numbers. Next they should incorporate the 10s beansticks to display numbers with two-digit place value. Three-digit numbers require the 100s flats. Example 1, below, depicts such place value representations.

3. Examples 2, 3, 4, and 5 cite methods for using the beans and beansticks to add, subtract, multiply, and divide. Please pay particular attention to the situations in which trading (sometimes called renaming, regrouping, borrowing, or carrying) is necessary. For these examples, students could work in pairs. If each student initially made ten beansticks and ten rafts, a pair of students will have enough to work these examples. Finally, students should keep a written record of the problems, processes, and outcomes.

Examples:

1. The numbers 3, 25, and 137 are displayed using beans and beansticks.

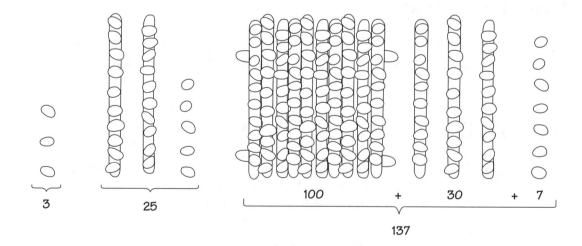

2. The problem $16 + 12 =$ ____ is solved below (in equation format) by simply combining the 1s beans and the 10s beansticks. In this case it is not necessary to trade.

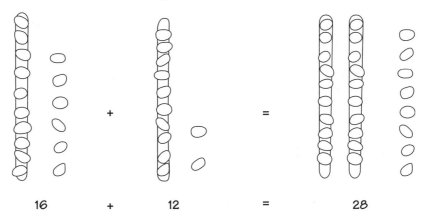

16   +   12   =   28

3. The problem $21 - 6 =$ ____ requires trading. Because 6 cannot be subtracted from 1, one of the 10s beansticks is traded for 10 single beans, allowing 6 to be taken away from 11. The player ends with one 10s beanstick and 5 single beans, or $10 + 5 = 15$.

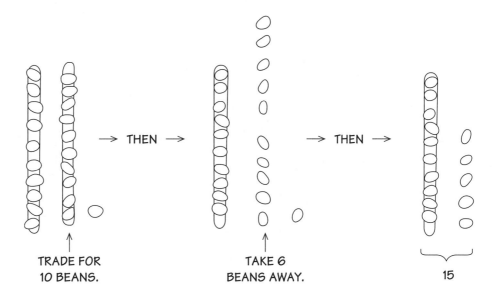

TRADE FOR   TAKE 6
10 BEANS.   BEANS AWAY.   15

4. Trading is necessary to solve the problem $3 \times 45 = \underline{\quad}$. Students should start by setting up this problem as an addition problem. Three 45s are set up using four beansticks and five loose beans. Notice that 10 of the 15 loose beans need to be traded for a 10s beanstick, and also that 10 of the beansticks are traded for a 100s flat.

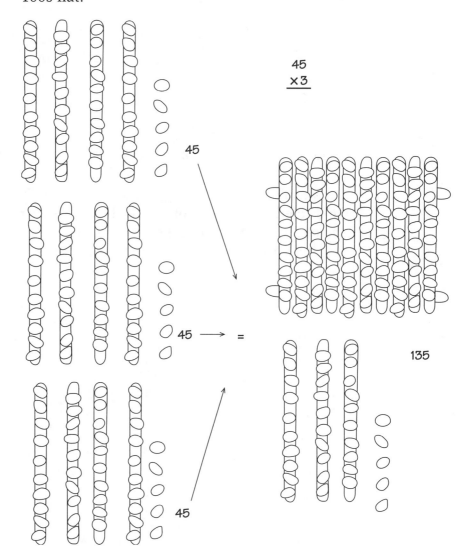

5. For the division problem $123 \div 27$, the student must figure out how many 27s are in 123. First 123 is represented using one raft, two beansticks, and three loose beans. Then the student represents the number 27 with two beansticks and seven loose beans. After careful counting and carrying, the student will determine that after displaying four 27s and adding them together, the result is not 123, but close. What is needed to make 123 is one beanstick and five loose beans, or 15. Because what is needed is less than 27, then this must be the remainder of the division problem. Therefore, the answer is 4 with a remainder of 15.

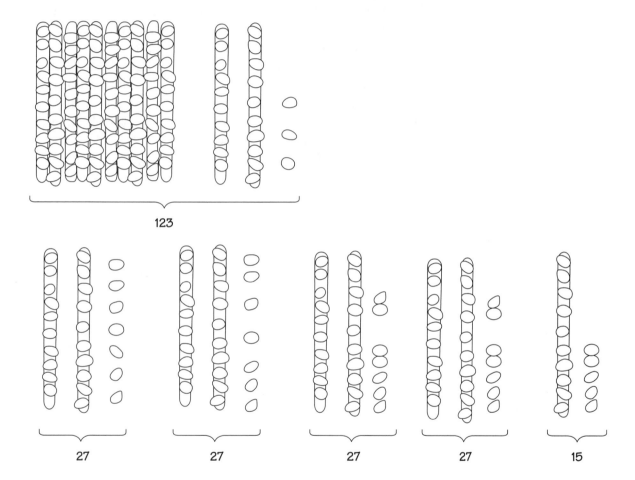

Extensions:

Beans and beansticks may be utilized with both larger numbers and decimals.

1. If students are working with numbers into the 1,000s, they can build these by stacking the 100s flats: to form each 1,000 requires that 10 of the 100s flats be piled together.

2. It can also be helpful to utilize visual representations in the bean-stick problems. For example, the number 253 might be quickly illustrated as shown below.

200   50   3

3. Beansticks might also be used (in reverse order) to portray decimals. For instance, if the 100s flat represents 1, then each 10s beanstick would equal .1 and each single bean would equal .01. Decimal computations can therefore be displayed. For example, adding .52 to .71 would entail adding five beansticks to seven beansticks to obtain one raft and two beansticks, then adding two single beans and one single bean to obtain three single beans. The answer would then be represented with one raft, two beansticks, and three single beans, which equals 1.23.

Chapter 7

Incredible Expressions

Grades K–8

☒ Total group activity
☒ Cooperative activity
☐ Independent activity
☐ Concrete/manipulative activity
☐ Visual/pictorial activity
☒ Abstract procedure

Why Do It:

Students will develop their number sense and make new mathematical connections.

You Will Need:

This activity can be done on the chalkboard or on a large piece of paper, with chalk or marking pens.

How To Do It:

In this activity, you or a student will specify a number for the whole class to represent in different ways. For example, 10 can be represented as $5 + 5$, $19 - 9$, $\sqrt{100}$, and so on. Keep a permanent record of the various ways of naming the same number by writing students' Incredible Expressions on a large piece of newsprint or butcher paper; or use the chalkboard to keep a short-term record.

Example:

The illustration below shows the Incredible Expressions for the number 21 that a group of students has devised.

$$3 \times 7 \qquad 10 + 10 + 1 \qquad \begin{array}{r} 7 \\ 7 \\ + \ 7 \\ \hline \end{array} \qquad 84 \div 4$$

$$22 - 1$$

$$1{,}000 - 979$$

$$\sqrt{441} \qquad (10 \times 2) + 1 \qquad \mathbf{100 - 79}$$

$$3 + 3 + 3 + 3 + 3 + 3 + 3$$

$$3^2 + 3^2 + 3 \qquad \mathbf{50 - 29}$$

Extensions:

Incredible Expressions may be simple addition or subtraction problems; or they may be more complex, involving multiple operations and exponents.

1. Each day, develop and list expressions to correspond with the calendar date. Students should also build these numbers with manipulatives (for example, 21 might be built with two bundles of 10 straws rubber banded together plus a single straw).

2. Restrict players to certain operations. For example, players might be directed to use only addition and subtraction.

3. Students who have had sufficient experience with the mathematical operations of addition, subtraction, multiplication, and division can be required to use all four of these operations. Advanced players might be instructed to find expressions that use parentheses,

exponents, square roots, fractions, decimals, percents, and so on. Be certain that they use the proper order of operations when computing an expression. The order is parentheses, exponents, multiplication and division (left to right), addition and subtraction (left to right). For example, the value of the expression $2(5 + 4)$ is different from the value of $2 \times 5 + 4$. The first expression simplifies to $2 \times 9 = 18$, and the second expression computes as $10 + 4 = 14$.

4. Older students can use calculators to create truly Incredible Expressions. If so, each player might be required to use at least five different numbers together with a minimum of three different operations.

Chapter 8

Number Cutouts

Grades K–8

☒ Total group activity
☒ Cooperative activity
☒ Independent activity
☒ Concrete/manipulative activity
☒ Visual/pictorial activity
☒ Abstract procedure

Why Do It:

This activity helps develop students' number sense and emphasizes mathematical connections.

You Will Need:

Graph paper (find square-inch and square-centimeter reproducibles on the following pages), scissors, glue or bulletin board pins, and a pencil are required.

How To Do It:

Cut the graph paper to make rectangular area cutouts for numbers from 1 to 100 (with each square on the graph paper representing the number 1) and label the cutouts accordingly (see Examples). Choose either square-inch or square-centimeter graph paper to complete the rectangular area cutouts for each number in as many ways as possible. Arrange them in order on a bulletin board or glue them to a large sheet of poster paper. Also, do as many of the Extensions as you can, and have students record their findings.

Examples:

Examples for the numbers 1 through 6 are shown below.

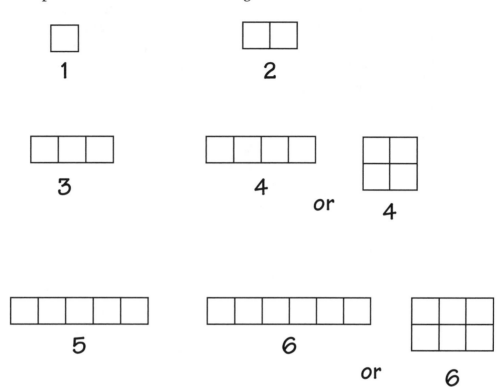

Extensions:

Use cutouts to help answer these questions.

1. Which of the number cutouts for Examples 1 through 6 can be shown in more than one way?

2. Which numbers from 1 through 6 might be called perfect square numbers? Why? Cut out three perfect square numbers that are not shown above. Students can also cut out the perfect square, such as the 2- by 2-unit square shown above, in a different color. This would make it stand out on the display and show that the number 4 is a perfect square number.

3. Show the number 12 with as many different rectangular area cutouts as possible. How about the number 16?

4. Discuss the idea of factors. For example, the number 12 has six factors: 1, 2, 3, 4, 6, and 12. How can you find the factors for each number from 1 to 100 by looking at your rectangular area cutouts?

5. Which numbers from 1 to 100 can be cut out in only one way? What are these numbers called? Students could also cut out these prime numbers using a different color, so that they stand out on the display. This could also lead to the definition of a prime number as a counting number with exactly two factors.

6. What do your cutouts show about multiplication? This question could lead to a discussion about the commutative property of multiplication, for example, $2 \times 3 = 3 \times 2$.

Number Cutout Graph Paper (cm)

        Making Sense of Numbers

Chapter 9

Celebrate 100 Days

Grades K–8

☒ Total group activity

☒ Cooperative activity

☒ Independent activity

☒ Concrete/manipulative activity

☒ Visual/pictorial activity

☒ Abstract procedure

Why Do It:

Students will increase their comprehension of number and numeral concepts for the numbers 1 through 100.

You Will Need:

For this ongoing activity, a roll of wide adding-machine paper tape; marking pens; 100 straws; rubber bands; and cans with 1, 10, and 100 written on them are needed. For the culmination activity, a wide variety of items that can be counted, separated, or marked into 100s will be needed. (*Note:* These items should be free or inexpensive, and many can be brought from home.)

How To Do It:

1. Begin the first day of school by taping a long length of paper tape to the classroom wall (perhaps affixed just above the chalkboard and spanning the width of the room). Since this is day 1, write a large numeral 1 at the left end of the tape and tell the students that a numeral will be added (for example, during the morning

opening exercises) for each day they are in school. Continue in this manner until day 5, at which time note that every fifth number will be circled. On the tenth day, note that every tenth number will have a square placed around it, pointing out that 10, 20, 30, and other 10s numbers have both circles and squares. This ongoing activity will continue until day 100.

2. Throughout the 100 days, use the cans marked 1, 10, and 100, along with straws and rubber bands, to help students understand the 1-to-1 connection with the numerals on the paper tape. On day 1, for example, both write the numeral 1 on the wall chart and have a student put 1 straw in the 1 can; on day 2, put a 2nd straw in the can. When day 10 arrives, put a rubber band around the 10 straws that have accumulated in the 1 can and move them to the 10 can, and so on.

3. When the class reaches day 100, it is time for a celebration! On that day it becomes each student's responsibility to do, make, count, separate, mark, or share 100 of something. It will certainly prove to be a fun learning experience. (See Example 2 and Extension 2 below for some possible ideas for types of 100s.)

Examples:

1. The illustration below shows the numerals on the wall tape and the straws in the number cans for the 23rd day of school.

2. These students are celebrating day 100 by showing 100 in their own ways!

Extensions:

1. Create *Incredible Expressions* (see p. 19), do a *Numbers to Words to Numbers* activity (see p. 71), or pursue *Calendar Math* (p. 50) to correspond to the day's number. Also consider making *Dot Paper Diagrams* (p. 112).

2. Have the students each collect 100 aluminum soda cans as part of a class project. Donate the recycling money to a charitable organization.

Chapter 10

Paper Plate Fractions

Grades 2–7

☒ Total group activity

☒ Cooperative activity

☒ Independent activity

☒ Concrete/manipulative activity

☒ Visual/pictorial activity

☒ Abstract procedure

Why Do It:

Students will gain visual and concrete exposure to basic fraction concepts. Advanced students may also use this activity to explore the concept of equivalent fractions.

You Will Need:

About 200 to 300 lightweight, multicolored paper plates are needed for a class of 30 students. Each student will need 6 plates, a pair of scissors, and a crayon or marking pen.

How To Do It:

1. Instruct the students to take a red plate and mark it 1 for one whole amount. Next have them fold or draw a line through the center of a blue plate and use scissors to cut along this line. Ask how many blue parts there are and whether they are equal. Because there are now two equal portions, students should write a 2 on each part and then a 1 above each 2 to indicate that each half is 1 of 2 equal parts. Note that we commonly call each part 1/2, but this really means 1 of 2 equal parts. Continue this process, with green plates being cut into 1/4s (1 of 4 equal parts), pink plates cut into 1/8s (1 of 8 equal parts), and so on.

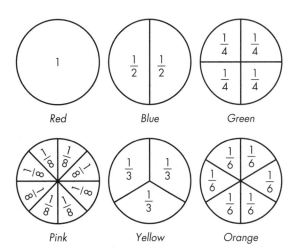

Red Blue Green

Pink Yellow Orange

2. When students have finished preparing the sections of plates, have them use their Paper Plate Fractions to explore equivalence concepts. Initially, direct them to use fraction pieces of the same sizes, to match these to the red "1 whole" plate, to keep records of what they find, and to share their findings with the whole group. Students will determine, for example, that $1/2 + 1/2 = 1$, and that $1/4 + 1/4 + 1/4 + 1/4 = 1$. After students have explored 1 whole, help them compare fractions to other fractions, again by creating physical matches. They will soon discover, for instance, that $1/4 + 1/4 = 1/2$ and $1/8 + 1/8 = 1/4$, but that $1/6$ and $1/4$ do not match up.

3. If students have mastered the preceding concepts, introduce them to a game called "Put Together One Whole." The students will need their paper plate fraction cutouts and a spinner (see below for illustration; see *Fairness at the County Fair*, p. 321, for directions on making a spinner). (A blank die can also be used with the faces labeled 1/2, 1/3, 1/6, 1/4, 1/8, and 1/8. Because there are six faces on a die, one of the fractions has to be used twice.) At each turn, players will spin the spinner, and using their red 1 whole paper plate as their individual game board may choose whether or not to lay the corresponding fractional part on top of the red plate. The object is to put together enough of the proper fractional pieces to equal exactly one whole. In the example noted below, the fractions spun so far are 1/4, 1/3, 1/8, 1/2, and 1/6. The first player opted not to use 1/3 or 1/6, and the second player did not use 1/8 or 1/2. Thus, in order for the game board to equal exactly one whole, the first player needs 1/8 and the second needs 1/4.

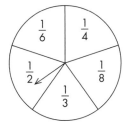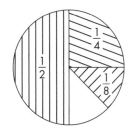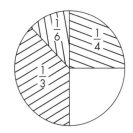

Example:

The students shown below are talking about their discoveries with 1/4 and 1/8, as these fractions relate to 1.

Extensions:

1. Expand the paper plate activity to give students experience with other common fractions, such as 1/9, 1/10, 1/12, and 1/16.

2. Advanced students might explore and show the meanings of such decimal fractions as .1, .5, .125, .05, and .25. These decimals correspond to 1/10, 1/2, 1/8, 1/20, and 1/4, and can be used to play the "Put Together One Whole" game explained above. Students will also see the relative sizes of these decimal numbers and be able to make comparisons.

Chapter 11

Bean Cups to 1,000

Grades 2–6

☒ Total group activity

☒ Cooperative activity

☒ Independent activity

☒ Concrete/manipulative activity

☒ Visual/pictorial activity

☒ Abstract procedure

Why Do It:

Students will experience number and place value concepts in concrete, visual, and abstract formats.

You Will Need:

A bag of dried beans (approximately 1 quart); 100 or more small cups (6-ounce, clear plastic cups are ideal; paper Dixie Cups will work); and a Place Value Workmat approximately 2 by 4 feet in size (sample shown here). If the activity is to be extended, about 100 blank 3- by 5-inch cards are needed, as well as marking pens to label the cards.

How To Do It:

1. Explain to the students that they will be counting sets of 10 beans and putting them in cups until they reach 1,000 beans (or more). Begin by spilling a quantity of beans onto the surface beside the Place Value Workmat. Have the players each put 10 beans in a cup and place all of their cups in the tens portion of the workmat. Continue by counting together by 10s to find the first 100, the second 100, and so on. Stack those sets of 10 cups to show 100s and place them in the hundreds segment of the workmat. Keep an ongoing record of

the numeral value of the beans counted. (For example, when each student in a class of 28 has filled a 10s cup, the total equals 280.) Continue in this manner until students have gathered 10 "bean cup" stacks of 100, and then move the ten 100s (equaling 1,000) to the thousands section of the Place Value Workmat. Note that the workmat is now displaying one group of 1,000 and zero groups of 100s, 10s, and 1s, for a total of 1,000. If players demonstrate sustained interest and more beans are available, the counting and recording may continue.

2. Following the initial group activity of gathering 1,000 beans, the students should work as individuals or in small groups to show a variety of numbers with bean cups on the Place Value Workmat. They should be instructed, for example, to build such numbers as 47, 320, 918, or 1,234. Also, one student may collect a number of beans and have another student determine the number and write it as a numeral. Similar visual- and abstract-level activities are noted in the Extensions below.

Example:

The players below have shown the number 1,325 with beans and bean cups.

Extensions:

1. Have two teams or individual players use two Place Value Workmats and a pair of dice to compete in "Get to 1,000 First." To start, one team rolls the dice and adds the two numbers together. If, for example, they role a 5 and a 6, they get to place a 10s cup of beans in the tens section and 1 bean in the ones portion. Then the other team takes a turn with the dice, recording the rolled number on their workmat. Each team, at their turn, adds to their

previously rolled totals. The first team to get to 1,000 (or any other preset number) wins the game.

2. Advanced players should also make use of visual- and abstract-level cards to do the activities above. Visual-level place value cards are shown in the figure above; abstract-level place value cards are shown in the figure below. For this extension, it will be necessary to trade cards: when students have accumulated ten of the 1s cards, they will trade these for one 10s card; and likewise when students have amassed ten of the 10s cards, they will trade them for one 100 card, and so on. Abstract-level cards showing 325 are illustrated on the workmat below.

Visual-Level Cards

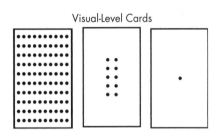

Abstract-Level Cards

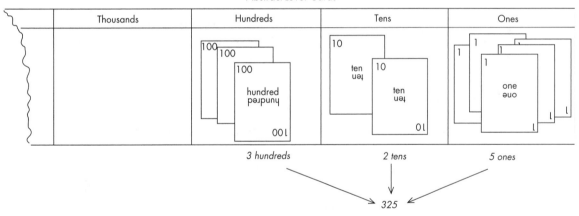

3. Other related activities are *Beans and Beansticks* (p. 13), *Celebrate 100 Days* (p. 27), and *A Million or More* (p. 62) in Section One, and *Dot Paper Diagrams* (p. 112) in Section Two.

4. The National Library of Virtual Manipulatives provides a great online resource for this activity. Visit the Web site www .nlvm.usu.edu and find the category "Number & Operations Grades 3–5." The Web site contains a chip abacus and base blocks that can be used for this activity.

Chapter 12

Dot Paper Fractions

Grades 2–8
☒ Total group activity
☒ Cooperative activity
☒ Independent activity
☐ Concrete/manipulative activity
☒ Visual/pictorial activity
☒ Abstract procedure

Why Do It:

Dot Paper Fractions will enhance students' comprehension of fractional parts and equivalent fractions.

You Will Need:

Photocopies of the "Dot Paper Fraction Problems" pages and the "4 by 4 Dot Paper Diagrams" page provided, colored markers or crayons for shading, and pencils to record results are needed. You can extend this activity by using larger dot diagrams from *Dot Paper Diagrams* (p. 112).

How To Do It:

Students will visually explore fractional parts and equivalent fractions using dot paper. There are three types of problems demonstrated in the Examples. After exploring the Examples with the class, have the students do the problems on the "Dot Paper Fraction Problems" handout. To extend this activity further, students can be provided with more copies of the "4 by 4 Dot Paper Diagrams" page and continue with the Extensions.

To begin, use the dot diagram sheet provided on an overhead projection system in order to go through each example with the class. Provide each student with a photocopy of the

"4 by 4 Dot Paper Diagrams" page, so they can follow along and try some sample problems on their own. After showing an example, give students a sample problem to do on their own. A sample problem for Examples 2 and 4 is provided. An important rule to tell students is that in order to draw a polygon (a closed, two-dimensional geometric figure with straight sides) on the dot diagram, the sides of the polygon must be formed by straight line segments, and a line segment is formed by connecting two dots. The line segment connecting two dots vertically or horizontally is considered one unit on the dot diagram.

After completing the three examples, have students work in pairs to solve the problems on the "Dot Paper Fraction Problems" handout. Answers to the handout are provided at the end of the Extensions section.

Examples:

1. Begin by displaying a rectangle that is 3 units by 4 units on the overhead. Discuss the different fractions that can be represented on this diagram by shading in a portion of the rectangle. Some fractions that should be discussed are shown in the figure below.

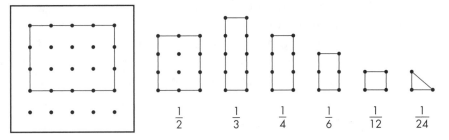

$$\frac{1}{2} \qquad \frac{1}{3} \qquad \frac{1}{4} \qquad \frac{1}{6} \qquad \frac{1}{12} \qquad \frac{1}{24}$$

2. At first explain in detail how one shaded portion is a fractional part of the whole and that one specific fraction (such as 1/2) is being represented. Then have students try to discover the fraction, given the shaded figure.

   Next, demonstrate that 1/3 = 2/6 = 4/12, as shown below. When students seem to understand, have them try to demonstrate the following sample problem on their "4 by 4 Dot Paper Diagrams" page.

   Sample Problem: Show that 1/2 = 3/6 = 12/24.

$$\frac{1}{3} = \frac{2}{6} = \frac{4}{12}$$

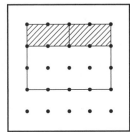 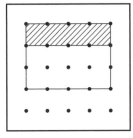 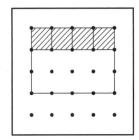

Dot Paper Fractions

3. Show a basic polygon, such as a rectangle, on the dot diagram. On another dot diagram, outline a smaller polygon that is a fractional part of the first polygon. Ask students what fractional part of the larger polygon the smaller polygon represents (see the figure below). Do this a few times, then ask students to make up their own problem. Then share some of the problems with the entire class and have them discuss the solutions.

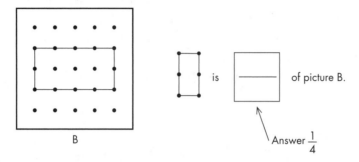

4. Outline any polygon on the dot diagram, and shade in a portion. Ask students what fractional part of the polygon is shaded. To demonstrate, divide the entire polygon into equal pieces, preferably using the shaded region as the guideline (see the figure below). Count the total number of equal sections and make that number the denominator of the fraction. Then count the number of shaded sections and make that number the numerator of the fraction. In some cases, you might have to simplify (reduce) the fraction.

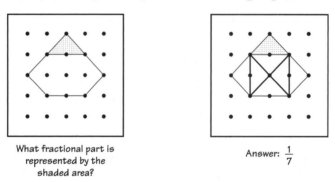

What fractional part is represented by the shaded area?

Answer: $\frac{1}{7}$

Have students draw and solve the following sample problem on their own. (Answer: 2/15).

Sample Problem:

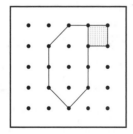

Extensions:

1. If you extend the size of the dot diagram, you can vary the size of the polygons according to the students' grade level. Larger and more complex polygons can be created at higher grade levels. (See more dot diagrams provided in *Dot Paper Diagrams*, p. 112.)

2. Students can work together in groups to make their own polygons that have a portion shaded (extending Example 4). They might also make up a problem for another group of students to solve.

3. Have students investigate all the different ways they can divide a square dot diagram of 4 by 4 units (or an expanded version) in half. Students will find many ways to do this, and the activity can lead to a discussion of area if students investigate whether the areas are the same.

Answers to Dot Paper Fraction Problems

1.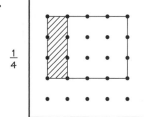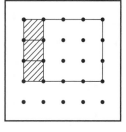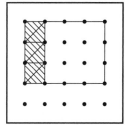

$\frac{1}{4}$ $= \frac{3}{12}$ $= \frac{6}{24}$

2. **a.** 1/2   **b.** 1/3   **c.** 1/18

   d. 1/8   **e.** 1/6   **f.** 1/3

3. **a.** 2/9   **b.** 2/12 = 1/6   **c.** 3/9 = 1/3   **d.** 1/6

   e. 2/5   **f.** 7/10   **g.** 8/14 = 4/7   **h.** 10/14 = 5/7

Dot Paper Fraction Problems

1. Show that 1/4 = 3/12 = 6/24.

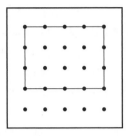 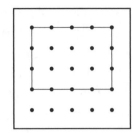 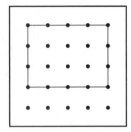

2. Using the figures below, answer the following problems in fraction form. For example, the figure on the left (problem a) is a fractional part of one of the pictures at the top labeled A, B, or C. Each problem refers to only one of the pictures at the top.

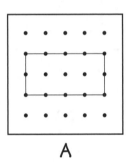 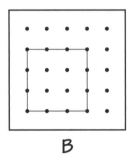 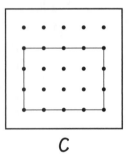

A   B   C

a. ⬜ is ⬜ of picture A

b. ▭ is ⬜ of picture C

c. ◿ is ⬜ of picture B

d. ⬜ is ⬜ of picture A

e. ▯ is ⬜ of picture C

f. ▯ is ⬜ of picture B

Copyright © 2010 by John Wiley & Sons, Inc.

More Dot Paper Fraction Problems

3. To the right of each figure, write the fraction that represents the part of the polygon that is shaded.

a.

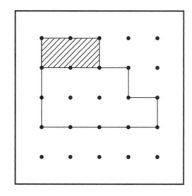

b.

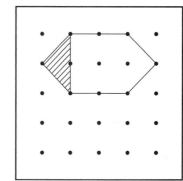

c.

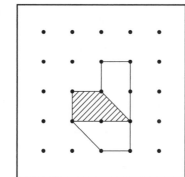

d.

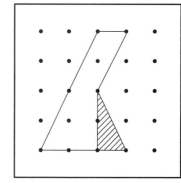

e.

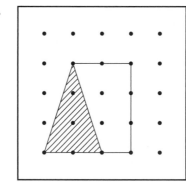

f.

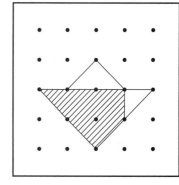

g.

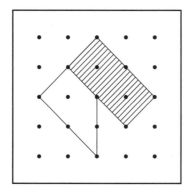

h.

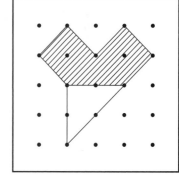

Chapter 13

Fraction Cover-Up or Un-Cover

Grades 3–8

☐ Total group activity
☒ Cooperative activity
☐ Independent activity
☒ Concrete/manipulative activity
☒ Visual/pictorial activity
☒ Abstract procedure

Why Do It:

This activity gives students conceptual experiences with fractions in a game-like setting.

You Will Need:

Each pair of students will need one die, stickers or tape, and construction paper or tagboard.

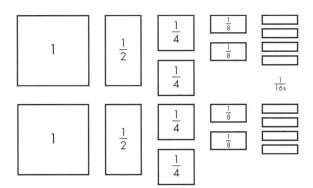

How To Do It:

1. The following activity is composed of two separate, but related, games. The following directions will help you initially set up the materials for the two games. Using stickers or pieces of tape, mark the sides of a die 1/4, 1/8, 1/16, and 1/16, leaving two blank sides. Each player should have his or her own game board made from construction paper or tagboard, and marked with the number 1. Any size game board is fine, as long as the fraction pieces cut correspond to the game board marked 1. Use the construction paper or tagboard also to make labeled fractional pieces, of varied sizes and colors, that correspond with the fractions on the die and that include a 1/2 piece (see above). Each player starts with a piece of construction paper or tagboard that is the size of the game board and divides it in half, labeling each piece with the fraction 1/2. Then they will do the same thing to make four pieces labeled 1/4 and so on. The fractional pieces from each player are placed nearby in a pile from which either player may draw, depending on their die rolls.

2. In "Cover-Up," the players each roll the die once to see who will begin, and the player with the greater fraction plays first. Player 1 rolls the die, and whatever fraction turns up determines the size of the piece he or she draws from the pile to cover a portion of his or her game board. Player 2 then rolls the die to find out how much of his or her game board to cover. Players 1 and 2 alternate rolling the die until one of them exactly covers his or her game board to equal 1. (*Note:* When rolling the die, if a blank turns up, it counts as zero, and no piece is put on the game board. Also, students may, and probably should, trade in combinations of smaller fraction pieces for larger ones as the game progresses. They might, for example, trade two 1/16 pieces for a 1/8 piece, or two 1/8 pieces for a 1/4 piece.)

3. "Un-Cover" is played in a reverse manner: the game board will be covered at the onset with fractional pieces, and the players will be removing them. Each player covers his or her game board with the 1/2, 1/4, 1/8, and two 1/16 pieces. Each time the player rolls the die, he or she may remove a fractional part from his or her game board. For example, if Player 1 rolls 1/8, he or she may remove either the 1/8 piece or two 1/16 pieces. (*Note:* As this game progresses, it will be necessary to trade pieces a number of times. If midway through the game, for example, Player 2 has only the 1/2 fraction piece on his or her game board and rolls 1/16, he or she must first trade in the 1/2 for 1/4, 1/8, and two 1/16 pieces before removing 1/16.) The game continues in this manner until one player has removed all of the fraction pieces from his or her game board.

Example:

The students shown below are playing "Cover-Up." Thus far Justin has rolled 1/16, 1/4, and 1/8. Currently Angelica has gotten two 1/4 pieces, which she has traded in for 1/2, and a 1/8 piece.

JUSTIN   ANJELICA

Extensions:

1. To enhance players' comprehension, change the shape of the game board and the fractional pieces. A few possibilities are shown below.

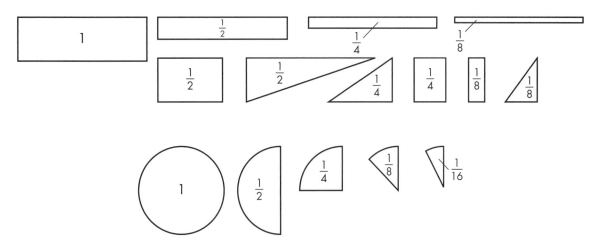

(*Note:* Encourage students to find as many possibilities as they can. For example, a student might use an octagon as a game board, and use fractional pieces such as 1/2, 1/8, 1/4, and 1/16.)

2. Players who are quite adept can play ''Cover-Up'' or ''Un-Cover'' with such fractions as 1/9, 1/10, 1/12, and 1/16, or even with 3/8 and 7/16. Challenge highly advanced students to role two dice at each turn. In this case they would be required to add (or subtract) two fractions, such as 1/12 and 3/8, and then determine exactly how much to ''Cover-Up'' or ''Un-Cover.''

3. Decimal fractions and percentages also work well with this activity. This will enable students to make the connections between numbers such as 1/4 and .25, or 1/8 and 12.5%, and to compare sizes of decimal numbers and percents.

Chapter 14

Post-it™ Mental Math

Grades 3–8

☒ Total group activity

☒ Cooperative activity

☐ Independent activity

☐ Concrete/manipulative activity

☐ Visual/pictorial activity

☒ Abstract procedure

Why Do It:

This activity enhances students' mental math skills, encourages the use and understanding of mathematical language, and stimulates logical thinking.

You Will Need:

Large-size Post-it notes (or index cards and masking tape) and a marking pen are required.

How To Do It:

1. Explain to the students that a number will be written on each of two Post-it notes and placed on the back of a chosen Post-it player, without that player being allowed to see them. The Post-it player must turn his or her back to the other group members, so that they may see the two written numerals, and then turn to face the group again. The group first needs to decide whether they want the Post-it player to say the individual numerals or the whole number (for example, if the player should say that the numerals are 4 and 9 or that the number is 49). One member of the group will then give the Post-it player a clue as to what the two numerals are or what the entire number is. The Post-it player then

makes a guess. If the guess is wrong, another member of the group will think of another clue, with the Post-it player guessing after each clue. When the Post-it player guesses the correct number, his or her score is the number of guesses it took to come up with the correct answer. Then the group chooses another player to wear new Post-it notes on his or her back, and the game continues. After all group members have had a chance to be the Post-it player, the player with the fewest guesses wins. The group can also have every member be a Post-it player more than once, the player with the smallest total number of guesses being the winner.

2. Clues can vary. Assume the two numerals are 4 and 9. A clue about the whole number might be that the number is greater than 20 but less than 60. A clue about the individual numerals might be that the difference of the digits is 5. Give younger children examples of clues on cards, such as "The difference of the digits is _?_," and have them fill in the blank before saying the clue to the Post-it player. Encourage students to use correct mathematical terms, such as *sum, difference, product, quotient, less than, greater than,* and so on.

3. To make the game more interesting, group members can choose to act out their clues. For example, the player might say, "I'm going to do five sit-ups, because the difference of the digits is five."Or a player might say, "We will run in place the product of the numerals," at which point group members run and count in unison, "One, two, three," all the way to thirty-six, for each time their left feet touch the ground.

Examples:

The group members shown above are younger students trying to get the Post-it player to name the two numerals (rather than the whole number) on his back.

The group members shown above are older students trying to get the Post-it player to say the number represented by the two numerals.

Extensions:

1. Beginners might work with a single numeral, and simply act out that number value. For example, a student might tap the floor ten times to try to get the Post-it player to guess 10. Later they can play "One More," in which the acted-out number value is one more than the number on the Post-it player's back. For example, if the group members complete six hops, the Post-it player guesses that 5 is the number on the Post-it. (Other options include "One Less," "Two More," and so on.)

2. Advanced players might work with three single-digit numerals on the Post-it player's back. If the chosen numerals were 4, 6, and 9, for example, the group players might be asked to subtract the smallest number from the largest and add the remaining number ($9 - 4 + 6 = 11$), before acting out the answer; or they might multiply the largest by the smallest and then divide by the middle-sized number ($9 \times 4 \div 6 = 6$).

Chapter 15

Calendar Math

Grades 3–8

☒ Total group activity
☒ Cooperative activity
☒ Independent activity
☐ Concrete/manipulative activity
☐ Visual/pictorial activity
☒ Abstract procedure

Why Do It:

Students will practice a variety of computational skills and discover interesting patterns.

You Will Need:

A calendar page or grid is required for each player or group, as are pencil and paper and colored pencils or crayons.

How To Do It:

Students will be using a calendar to find numbers to perform computations, to see patterns, and to play a game. Tear a page from a calendar or have students make their own pages from a 5 by 7 grid made up of 35 squares. (If students make their own pages, label the seven columns Sunday, Monday, Tuesday, Wednesday, Thursday, Friday, Saturday and number the squares beginning with the first day as 1 and ending with the last day as 28, 29, 30, or 31, depending on the number of days in the current month.) Have students do some investigating, beginning with the questions in the Examples and Extensions sections (you can also come up with your own questions), and then see what else students can discover on their own. Have them share findings with some of the other students.

Sunday	Monday	Tuesday	Wednesday	Thursday	Friday	Saturday

Examples:

Ask students to attempt the following problems.

1. Add the dates for the first two Tuesdays together and get ____.
2. Subtract the date for the second Friday from that of the third and get ____.
3. All of the Monday dates added together equal ____.
4. Subtract the date of the first Wednesday from the date of the third Thursday and you get ____.
5. Multiply the date of the second Wednesday by that of the third Wednesday and get ____.
6. Color some patterns on your calendar. For example, count off by 2s (2, 4, 6, 8, and so on) and share the pattern you find. Use a different color and mark in the multiples of 3. Were any of the numbers marked as both 2s and 3s? Why did that happen? Try coloring in some other number multiples and share what happens.

Calendar Math

Extension:

Play "Corner to Corner" with one, two, or three other players. You will need a large calendar page, different colored markers, and about forty small cards marked as shown below.

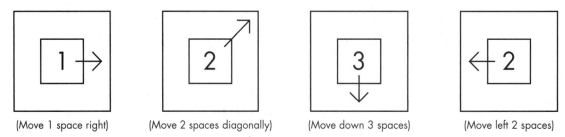

(Move 1 space right)   (Move 2 spaces diagonally)   (Move down 3 spaces)   (Move left 2 spaces)

Shuffle the cards and pile them upside down next to the calendar page. Have each player put his or her marker on a calendar corner number. Also decide who will go first, second, third, and fourth. On their individual turns, players will select one of the cards and devise and solve a math problem using the two calendar numbers (the number on which their markers began and the one to which their cards directed them). After a player has picked a card, devised a problem, and solved it correctly, he or she can move his or her marker to the spot indicated by the card drawn. For example, on the calendar shown below, Ryan's red marker is on the 5, and the card he drew has indicated that he must

move down two spaces to the 19. Next he needs to decide on a math problem using those numbers, perhaps $19 - 5 = 14$. If the other players agree that he is correct, Ryan gets to move his marker to the 19. Play continues in this manner. If a player is at the bottom of a column and the card tells him or her to move down three spaces, the player goes back up to the top of the column and moves down three from there. If a player is at the end of a row and gets a card that will take him or her off the board, that player should wrap around to the other end of the row and move from there. Finally, if a player is at a corner and gets a card to move diagonally off the board, he or she should go to the opposite corner and move diagonally from there. The first player to reach the corner opposite to his or her starting place is the winner.

Chapter 16

Let's Have Order

Grades 2–8
- ☒ Total group activity
- ☒ Cooperative activity
- ☐ Independent activity
- ☐ Concrete/manipulative activity
- ☐ Visual/pictorial activity
- ☒ Abstract procedure

Why Do It:

This activity helps reinforce students' understanding of place value and number order.

You Will Need:

A set of large cards (8-1/2 by 11 inches, the size of a standard sheet of paper) is required.

How To Do It:

Form teams with ten players, and number the cards 0 through 9 for each team (it may be desirable to have duplicates of the numbers for each group). Pass out the cards (unless players have made their own) and put one person in charge of each numeral. Also designate locations where each team's numeral-carrying players are to report and display their numbers in the correct order. (Chalkboard locations marked One, Ten, Hundred, and so on work well.) Students from each team will be reporting to their location, putting themselves in the order needed to display the number the leader calls out.

To begin, the leader calls out a number, perhaps "Four hundred two." After teams hold a brief conference, the leader says, "Let's have order!" The team members with the numerals 4, 0, and 2 move directly to their teams' designated locations and display the numeral cards in their proper order. Each team displaying 4 in the Hundred position, 0 in the Ten, and 2 in the One score 3 points (1 point for each numeral in the proper place). For greater competition, the team that first displays the right answer can receive a bonus point.

Example:

The number called out for the teams depicted below is 402.

INCORRECT RESPONSE
(THIS TEAM SCORES 1 POINT)

CORRECT RESPONSE
(THIS TEAM SCORES 3 POINTS)

Extensions:

1. Call out numbers in place value terms only, and have the players show the numerals and state the standard number names. The leader, for example, might call for five thousands, six hundreds, three tens, and zero ones. The players would then have to say, "Five thousand six hundred thirty."

2. If the leader calls out a number with duplicate numerals, allow the player in charge of the repeated numeral to decide whether he or she can "reach" far enough: in other words, for a number like 303, the player in charge of 3s could stand behind the 0s player and reach a 3 out to either side of her or him. However, for a number like 25,432, the 2s player would need to put another team member in charge of one of the 2s.

3. The leader could call out some numerals, such as 5, 3, and 7, and the players would come up to the board to make the smallest or largest number possible with these numerals. Teams would not only receive points for having numerals in the correct place but also gain a point for calling out the correct name for their number, even if the number is not the correct answer for finding the smallest or largest. For example, the smallest number using 5, 3, and 7 would be 357, and the team would have to say, ''Three hundred fifty seven.''

4. Ask advanced players to deal with large numbers, perhaps into the billions or more. For example, how might three hundred twenty billion, ninety-one million, sixty-two thousand, seven be shown? (The answer is 320,091,062,007.)

Chapter 17

Reject a Digit

Grades 3–8

☒ Total group activity
☒ Cooperative activity
☐ Independent activity
☐ Concrete/manipulative activity
☐ Visual/pictorial activity
☒ Abstract procedure

Why Do It:

Students will engage in logical thinking as they discuss place value as it relates to computations involving addition, subtraction, multiplication, and division.

You Will Need:

A duplicate of the spinner below and one paper clip are required for each group of players, along with pencils and paper. One paper plate per group is optional.

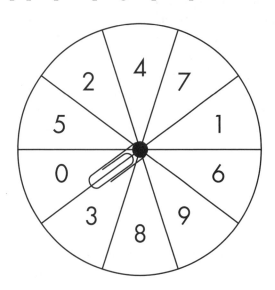

How To Do It:

1. Before beginning this activity, each group of students should make a spinner. An enlarged copy of the spinner shown numbered 0 through 9 should be provided to each group, and students can cut it out and glue it to a paper plate, if more stability is desired. To use the spinner, the student lays the paper clip on its side so that one end overlaps with the center point, and puts a pencil through this end loop of the paper clip such that the point of the pencil is on the center of the circle. The student then uses his or her other hand to flip the paper clip. The paper clip will randomly point to different numbers.

2. Students first need to draw any number of digit boxes and one reject box (see Example) on their individual record-keeping pages (or one record sheet can be made and photocopies provided). Students in a group will take turns spinning the spinner. The number of total spins will be one more than the total number of digit boxes on the record sheet, to allow for a reject digit. The description below will describe the rules for a game.

   When first introducing "Reject a Digit," have students attempt to get the largest (or smallest) numeral. In the Example, the students have each attempted to get the largest numeral containing 1s, 10s, 100s, and 1,000s. The rules they followed were: (1) each time the spinner stops, all students are required to write the number that comes up in a digit box or in the reject box; (2) only one numeral can be written in the reject box; (3) once a numeral has been written it cannot be moved; (4) the spinner is spun once more than the number of digit boxes, which allows one rejection.

3. Once students are familiar with the basic procedure, "Reject a Digit" may be played in a wide variety of computation situations (see the Extensions for suggestions). An important process in this activity is the discussion students have about their outcomes. Allow students to comment on what happened, what worked or did not work, why they think their choices resulted in certain outcomes, what might have been done differently, and how a change in rules might affect the outcomes. In this way, students will gain experience not only with place value and computation concepts but also with logical thinking.

Example:

In the game of "Reject a Digit" shown below, the students had each been trying to get the largest numeral. The spinner stopped in succession on 8, 4, 9, 3, and 2. Dan won, of course, but the discussion to follow is likely more important; of concern might be what numbers were achieved, why a number was put in a certain place value location or was rejected, and so on.

Extensions:

1. Do the activity using addition, subtraction, multiplication, or division. In each of the situations below, it is necessary to spin for six numerals; five are to be placed in digit boxes, and one is to be rejected. Before starting, each group should be told whether they are looking for the largest or smallest solution, or even the solution closest to a target number. As the students proceed, entering numbers in boxes, they should keep in mind whether they are trying to get the largest or smallest sum, difference, product, or quotient. After all numbers are placed into boxes, each student should perform the computation and compare answers.

☐ reject   ☐ reject   ☐ reject   ☐ reject

$$\begin{array}{r}\square\square\square\\+\ \square\square\end{array}$$ $$\begin{array}{r}\square\square\square\\-\ \square\square\end{array}$$ $$\begin{array}{r}\square\square\square\\\times\ \square\square\end{array}$$ $$\square\square\,)\overline{\square\square\square}$$

2. Spin for a selected number of digits (perhaps four) and challenge the players to use them to devise all possible problems of a certain type, such as addition. For example, if the numbers happen to be 1, 2, 3, and 4, some possible arrangements might include $1 + 2 + 3 + 4 = $ ____, $12 + 34 = $ ____, $31 + 42 = $ ____, $43 + 21 = $ ____, $1 + 23 + 4 = $ ____, and $123 + 4 = $ ____.

3. Calculators may prove helpful, especially for problems with large numbers of digits. Also, when players are able, consider problems with more than eight digits (most calculators have only an eight-digit display).

4. Another option is to extend this activity to include fractions. For "Fraction Reject a Digit," students can use the spinner numbered 0 through 9 or a standard die. If the number 0 shows up on the spinner, it cannot be played in the denominator of a fraction because you cannot divide by zero. Shown on page 61 are three possible options for game boards. Either have students draw the different game boards or provide photocopies of one of the game boards shown. The game proceeds as follows: (1) before the game starts, the leader decides whether the players will be playing for the smallest sum or largest sum; (2) a player spins the spinner or rolls the die and calls out the number; (3) when a number is called out, each player is required to write the number in one of the shapes on the game board. For example, a numeral could be written in all the squares, or in all the circles, or in all the triangles, or in the diamond, which indicates that the numeral has been rejected; (4) once a numeral is written it cannot be moved; (5) this process continues until a total of four numbers have been called out; (6) each player then computes his or her sum, and the player with the smallest or largest sum is the winner.

GAME A

$$\frac{\bigcirc}{\square} = \underline{\quad\quad}$$

$$\frac{\triangle}{\square} = \underline{\quad\quad}$$

$$\frac{1}{\square} = \underline{\quad\quad}$$

TOTAL = \underline{\quad\quad}

 REJECT

GAME B

$$\square + \frac{1}{\triangle} = \underline{\quad\quad}$$

$$\bigcirc + \frac{1}{\triangle} = \underline{\quad\quad}$$

$$\triangle + \frac{1}{\triangle} = \underline{\quad\quad}$$

TOTAL = \underline{\quad}

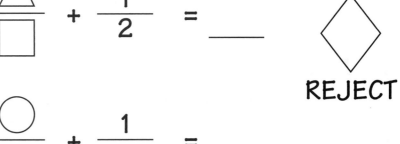 REJECT

GAME C

$$\frac{\triangle}{\square} + \frac{1}{2} = \underline{\quad}$$

$$\frac{\bigcirc}{\square} + \frac{1}{4} = \underline{\quad}$$

REJECT

TOTAL = \underline{\quad}

Reject a Digit

Chapter 18

A Million or More

Grades 4–8

☒ Total group activity
☒ Cooperative activity
☐ Independent activity
☐ Concrete/manipulative activity
☒ Visual/pictorial activity
☒ Abstract procedure

Why Do It:

Students will visually conceptualize how many a million is and further develop place value concepts, as these relate to how many thousands and hundreds are in a million.

You Will Need:

This activity requires dot paper sets for 10,000 (see reproducible version at the end of this activity); 1,000; 100; plus possibly 10 and 1 (it is sometimes easier to have the players draw 1- and 10-dot amounts themselves), and pens or pencils.

How To Do It:

1. Begin by examining and comparing 1 dot, a 10-dot strip, and a 100-dot square. Then bring out the 1,000-dot strips and ask how many of the 100-dot squares would make the same amount; also ask how many of the 10-dot strips would equal the 1,000-dot strip. Make the same types of comparisons with a 10,000-dot page. Now have the players figure out how many of the 10,000-dot pages it will take to make a million. Once they have determined the number of pages needed, have them make a million by posting 100 such pages

side by side on a large bulletin board or on butcher paper (if possible, the best arrangement is in a single row).

2. Once the Million Bulletin Board is in place, have the players find exact dot locations for numbers between 1 and 1,000,000. When doing so, be certain to establish a precise counting and locating procedure; the players should always begin at the lower left-hand corner and count up to 10, then go to the bottom of the next column and proceed from 11 to 20, and so on, until the first 100-dot square has been completed. If students are seeking a number greater than 100, the player may count squares of 100 from the bottom up, keeping in mind that each full column should equal 1,000 and each full page should equal 10,000.

Example:

The dot diagram shown displays the number 1,234.

Extensions:

Once the players have grasped the basic number and place value concepts for numbers up to a million or greater, it is fun to extend those ideas to everyday life experiences.

1. A whole class activity, or even one for the entire school, is to have the players bring "statistics" and place them on the Million Bulletin Board. They must clip a statistic or number data relating to a number between 1 and 1,000,000 from a newspaper or magazine, or obtain such information from radio or television, and bring it to school. The data clipping must then be glued to an index card or, if derived from radio or TV, transcribed and written on an index card, along with the inscription "Located by (Student's Name)." Then place the information card on the bulletin board and use push pins and yarn to connect it with the exact dot. For instance, a student might bring data that show an automobile priced at $19,995, the average price of a home in a certain area to be $189,623, or a typical teacher salary to be ???

2. Include space on the Million Bulletin Board (probably to the far right) for a More Than a Million category. On it the students can place data cards with larger amounts—such items as the number of albums sold by popular music stars or the salaries of some professional athletes.

3. Also included on the Million Bulletin Board (probably to the far left at the bottom) could be a Less Than One category. In it would be data relating to fractions or decimals between 0 and 1 (for example, a 1/2 price sale, or that a soda can contains .354 liters) or even negative numbers (for example, it was −20 degrees F° in Nome, Alaska).

10,000 Dots

A Million or More

Chapter 19

Smallest and Largest

Grades 4–8
- ☒ Total group activity
- ☒ Cooperative activity
- ☒ Independent activity
- ☐ Concrete/manipulative activity
- ☐ Visual/pictorial activity
- ☒ Abstract procedure

Why Do It:

Students will enhance their understanding of estimation, bolster their mental-math and logical-thinking skills, and gain computation practice.

You Will Need:

Number charts (see the reproducible handouts), need to be duplicated for the players. However, if played as a group activity, a single copy of the appropriate chart, displayed with an overhead projector or on the chalkboard, will suffice.

How To Do It:

In this activity, students will be filling in the "Smallest and Largest" charts in pairs. Each group will fill in each row as they discuss the possibilities for the first and second number. When first introducing an activity involving a "Smallest and Largest" chart, note any game rules (such as no zeroes allowed), demonstrate how to complete one or two problems, and discuss reasonable estimates, mental-math procedures, and why certain numbers were chosen. In the Example below, zeroes were allowed, but not in the first digit of a number. After completing an easy problem and a more difficult one together, the students should individually, or in small groups, try a problem on their own. The procedures they used and the solutions they decided on should be discussed, and brief notes should be recorded by each group. Finally, when the students have successfully completed the listed problem situation, they should devise some problems of their own, record their findings (in the rows at the end of the chart), and share their procedures and outcomes with the entire group.

Example:

Students have dealt with two of the problem situations on the following "Addition—Smallest and Largest" chart. Most players were able to master the problem situation in the first row quite easily, but the second problem required more analysis for varied reasons. (For example, some players thought, when seeking the smallest sum, that the first number should be 111 and the second 1,111.) The discussion is perhaps the most important aspect of the activity.

ADDITION – SMALLEST AND LARGEST

Number of Digits in First Number	Number of Digits in Second Number	Estimated Smallest Sum	Actual Smallest Sum (Show Work)	Estimated Largest Sum	Actual Largest Sum (Show Work)	My Notes
1	1	2	1+1=2	18	9 + 9 = 18	*My partner and I agree.*
2	2					
2	3					
3	3					
3	4	1200	100 +1000 1100	11,000	999 +9,999 10,998	*At first I forgot about using zero.*

Extensions:

1. Do the activity for subtraction using the "Subtraction—Smallest and Largest" chart. Unless the players are to deal with negative answers, be certain that the first number has more (or larger) digits than the second.

2. Do the activity for multiplication and division using the charts provided. To avoid confusion when students are working with division, designate the first number as the dividend and the second as the divisor.

3. The activity might be made more challenging by restricting the digits that can be used; for example, allowing only 6s, 7s, and 8s in certain cases.

4. Calculators may prove helpful, especially for problems with large numbers of digits. Also, for able players, consider problems with more than eight digits (most calculators have only an eight-digit display).

ADDITION — SMALLEST AND LARGEST

Number of Digits in First Number	Number of Digits in Second Number	Estimated Smallest Sum	Actual Smallest Sum (Show Work)	Estimated Largest Sum	Actual Largest Sum (Show Work)	My Notes
1	1					
2	2					
2	3					
3	3					
3	4					

SUBTRACTION — SMALLEST AND LARGEST

Number of Digits in First Number	Number of Digits in Second Number	Estimated Smallest Difference	Actual Smallest Difference (Show Work)	Estimated Largest Difference	Actual Largest Difference (Show Work)	My Notes
2	1					
3	2					
4	2					
4	3					
5	4					

MULTIPLICATION — SMALLEST AND LARGEST

Number of Digits in First Number	Number of Digits in Second Number	Estimated Smallest Product	Actual Smallest Product (Show Work)	Estimated Largest Product	Actual Largest Product (Show Work)	My Notes
1	1					
2	2					
3	2					
3	3					
4	3					

DIVISION — SMALLEST AND LARGEST

Number of Digits in Dividend	Number of Digits in Divisor	Estimated Smallest Quotient	Actual Smallest Quotient (Show Work)	Estimated Largest Quotient	Actual Largest Quotient (Show Work)	My Notes
2	1					
3	1					
3	2					
4	2					
4	3					

Making Sense of Numbers

Chapter 20

Numbers to Words to Numbers

Grades 4–8

☒ Total group activity
☒ Cooperative activity
☒ Independent activity
☐ Concrete/manipulative activity
☐ Visual/pictorial activity
☒ Abstract procedure

Why Do It:

Students will practice writing numbers as both numerals and words, and will compare their results.

You Will Need:

Chalkboard and chalk—or an overhead projector and pens—are required, as well as pencils and paper.

How To Do It:

This activity should first be played by the entire class. Begin by asking the students to select a number. Once decided, everyone should write it in numeral form on his or her papers and underline it; you should do the same on the chalkboard (see Example 1, in which 63 was chosen). Next, instruct the students to write that number as a word. You will again follow by writing the word on the chalkboard. Instruct the students to count the letters in that word, record the number, and then write the word for the new number. You and the class continue this procedure until the number word "four" continually repeats, as demonstrated in the Examples.

Examples:

1. The number in the first column shown below is 63.

2. The number chosen for the second column is 157.

1. Number picked:	**63**	
2. Written form:	sixty-three	
3. Counted letters:	10	
4. Written form:	ten	
5. Counted letters:	3	
6. Written form:	three	
7. Counted letters:	5	
8. Written form:	five	
9. Counted letters:	4	
10. Written form:	four	
11. Counted letters:	4	
12. Written form:	four	

←(NOTE: About here the players will realize that 4 will continue to repeat.)

1. Number picked:	**157**	
2. Written form:	one hundred fifty-seven	
3. Counted letters:	20	
4. Written form:	twenty	
5. Counted letters:	6	
6. Written form:	six	
7. Counted letters:	3	
8. Written form:	three	
9. Counted letters:	5	
10. Written form:	five	
11. Counted letters:	4	
12. Written form:	four	

3. In the situation below, one of the players has made an error when spelling the number word. The number chosen was 45. When this situation arises in the classroom, you could pair up two students who have different outcomes, and the students could find the error.

Player A

1. Number picked: 45
2. Written form: fourty-five
3. Counted letters: 10
4. Written form: ten
5. Counted letters: 3
6. Written form: three
7. Counted letters: 5
8. Written form: five
9. Counted letters: 4
10. Written form: four

Player B

1. Number picked: 45
2. Written form: forty-five
3. Counted letters: 9
4. Written form: nine
5. Counted letters: 4
6. Written form: four

Extensions:

Utilize the *Numbers to Words to Numbers* process for practice in several formats and at a variety of academic levels.

1. If you are working with primary students, you might want to practice with numbers no more than 20. The students may also need to follow you a number of times as you go through the process at the chalkboard or on the overhead projector.

2. When they are familiar with the procedure, the students may practice and check their work in pairs or cooperative groups. Each individual (or group) should work independently with the selected number and then compare outcomes with others.

3. Advanced players can try more complex numbers. For example, they might try 1,672,431, which in written form is one million, six hundred seventy-two thousand, four hundred thirty-one. (*Hint:* Remember to use ''and'' only to denote a decimal point.)

Chapter 21

Target a Number

Grades 4–8

☒ Total group activity
☒ Cooperative activity
☒ Independent activity
☐ Concrete/manipulative activity
☐ Visual/pictorial activity
☒ Abstract procedure

Why Do It:

This activity will reinforce students' understanding of place value, as well as their computation, reasoning, and communication skills.

You Will Need:

One die or spinner and a pencil are required. If students are working on a chalkboard or whiteboard, then chalk or whiteboard pens are also needed.

How To Do It:

1. In this activity, students will begin by drawing shapes in a predetermined arrangement. You will select an operation, and the students will place numbers in the shapes so that when the computation is complete they are close to a target number.

   Begin by selecting geometric shapes, such as

Then decide which operation will be used (addition, subtraction, multiplication, or division). Each student then decides individually where to place his or her shapes within an arrangement you have specified (see the Example below). You next select a target number (any number that could be an answer to the problem set up by any student.)

2. Now select the first shape to be considered and roll a die (or use a spinner) to determine the number to be placed in that shape. Then choose another shape and roll or spin for a number; the students place the number in that shape. Play continues in the same manner for the remaining shapes. When all the shapes are numbered, the students use the specified operation and complete their computations. Have the class discuss the varied problems and solutions they have found. The student or students who achieve or are closest to the target number win the round.

Example:

$3 = \bigcirc$, $1 = \square$, $4 = \triangle$, and $6 = \diamondsuit$. The preceding numbers were rolled in order and matched with the specified shapes. The target number was 850, and the operation was multiplication. The problems and solutions determined by three different players are shown below.

Extensions:

1. Use only a few geometric shapes or limit the operations (perhaps to only addition or subtraction) if you wish the games to be quite easy. For more complex games, increase the number of shapes utilized.

2. Allow the students to save the numbers until all have been rolled. Then let them individually arrange their numbers to see if they can "hit" the target number!

3. Have the students place their numbers as rolled, but allow them to add, subtract, multiply, or divide as an individual choice.

4. Use the *Target a Number* procedure with fraction operations, such as $\dfrac{\bigcirc}{\square} \times \dfrac{\Diamond}{\triangle} =$

5. Students could also try using parentheses and brackets, such as $\bigcirc\triangle + (\square \times \Diamond) - (\triangle \div \bigcirc) =$

Chapter 22

Fraction Codes

Grades 4–8

☒ Total group activity

☒ Cooperative activity

☒ Independent activity

☐ Concrete/manipulative activity

☒ Visual/pictorial activity

☒ Abstract procedure

Why Do It:

This activity enhances students' conceptual understanding of fractions (or percents or decimals) through the use of codes.

You Will Need:

Students each will need a prepared Fraction Code Message (one example is included here), as well as a pencil or pen.

How To Do It:

The first time students attempt to decipher Fraction Codes, provide them with a prepared code message (see Example) that they must solve. They may work independently or in cooperative groups as they try to determine the message from such clues as being asked to use the first 1/3 of the word *frenzy*, the first 3/8 of *actually*, and the last 1/2 of *motion* to form a word (fr + act + ion = fraction). After working with several sample coded messages, they may devise some of their own (see Extensions).

Example:

The students are asked to solve the "Fractions and Smiles" code below. The first two lines are already solved for them.

FRACTIONS AND SMILES

Last 1/2 of take	*ke*	First 1/4 of opposite	____
Last 2/5 of sleep	*ep*	Last 1/3 of stable	____
First 3/5 of smirk	____	First 2/3 of wonderful	____
First 1/4 of leap	____	First 3/5 of whale	____
Last 3/5 of being	____	Last 1/5 of generosity	____
First 1/2 of item	____	First 2/5 of ought	____
First 1/3 of matter	____	Last 3/4 of care	____
First 1/4 of keep	____	First 1/3 of use	____
First 1/5 of especially	____	Last 1/3 of abrupt	____
First 1/10 of perimeters	____	Last 1/2 of do	____
First 1/10 of equivalent	____		

The message is: *Keep smiling; it makes people wonder what you are up to.*

Extensions:

1. To expand players' understanding, devise coded messages that must be solved using percentages or decimals. For example, students might decipher a breakfast food from such clues as being asked to use the first 50% of the word *chip*, the middle 33-1/3% of *cheese*, the final 25% of *poor*, the first 40% of *ionic*, and the first 25% of *step* (ch + ee + r + io + s = Cheerios).

2. Challenge the students, if they are able, to devise their own Fraction or Decimal Codes. Have them use spelling or vocabulary words as part of their codes, and also encourage them to use mathematical words.

3. Students could also be asked to perform an operation with fractions to discover the fractional part of the word they are looking for, as will be the case when they are working with "A Good Rule" on the next page (Answer: *Perform an act of kindness today*). Remind players that all fractions should be simplified (reduced) before finding the part of the code.

A Good Rule

First 1/7 + 2/7 of percent ____   Second 7/8 − 4/8 of cylinder ____

First 1/14 + 1/2 of formula ____   Last 7/12 − 1/4 of
                                     completeness ____

First 2/3 × 3/5 of angle ____   First 1/4 + 3/20 of total ____

Middle 3/4 × 4/7 of fractal ____   First 8/12 × 3/4 of data ____

Last 3/10 + 1/10 of proof ____   Last 3/4 ÷ 6 of geometry ____

First 2/5 ÷ 8/5 of kite ____

The "Good Rule" is:

_____

_____ .

Chapter 23

Comparing Fractions, Decimals, and Percents

Grades 4–8

☒ Total group activity

☒ Cooperative activity

☐ Independent activity

☒ Concrete/manipulative activity

☒ Visual/pictorial activity

☒ Abstract procedure

Why Do It:

Students will understand and compare the relationships between fractions (and the division problem they represent), decimals, percents, and a variety of applications of each.

You Will Need:

This activity requires a large roll of paper (2 to 3 feet wide and perhaps as long as the classroom), marking pens of different colors, a yardstick or meter stick, string, scissors, glue, and magazines that may be cut up.

How To Do It:

1. Students will be drawing a chart designed to compare fractions to decimals and percents. The chart will have a vertical axis labeled 0 to 1 (to start) and a horizontal axis labeled with some different ways a fraction between 0 and 1 could be represented.

To begin, roll out several feet of the paper on a flat surface. Have students use the pens and yardstick to draw a vertical and horizontal axis and then several vertical number lines about a foot apart (see Example). On the vertical axis, have students write 0 at the bottom and 1 at the top. Then they will determine and mark in the fractions with which they are familiar. One way to do this is to cut a piece of string the length of the distance between the 0 and 1 and have the students fold it in half to help locate and mark the 1/2 position; then fold it in fourths to determine 1/4, 2/4, 3/4, and so on. Though the chart may get a bit cluttered, have the students position and mark on the number line as many fractions as possible. Also, be certain to discuss the meaning of each fraction and its relative position, dealing in particular with such queries as "Why is 5/8 between 1/2 and 3/4 on the number line?"

2. Have the students label the first vertical line to the right of the vertical axis "division meaning." Then, for each of the listed fractions, they should write the division problem represented, making sure it is directly across from the corresponding fraction. For example, 3/4 can be read as 3 divided by 4 and written as $4\overline{)3}$. Then, on the next vertical line, have the students compute the division problem (possibly using a calculator) and list the decimal representation.

3. The third vertical line to the right of the vertical axis might be used to make comparisons to cents (¢) in a dollar. Again using 3/4 as an example, 3/4 of a dollar can be written as $.75 or 75¢. In regard to the next vertical line, ask, "How many cents are there in one dollar? If 3/4 of a dollar is 75¢, how might this be written in terms of 100¢?" The response should be recorded as 75/100. This leads naturally to the next vertical line, on which students can derive percent (meaning per 100); the 75/100 translates easily to 75%.

4. Another vertical line might depict a visual representation or practical use of the fraction, decimal, or percent. For example, a picture of 3/4, .75, or 75% of a pizza might be cut out of a magazine and pasted onto the number line. Another example would be to portray a fraction, decimal, or percent of a group. If 8 elephants were pictured, for instance, the students might draw a fence around 6 of them to show 3/4, .75, or 75% of the elephant herd.

5. Finally, have students draw and mark subsequent vertical lines, based on either their interests or the need to develop concepts further. For example, a number line related to time, labeled "... of an Hour," might include how many minutes make up a given fraction of an hour (for example, 2/3 of an hour is 40 minutes). Each of the vertical lines should, in time, be fully filled in to correspond with the fractions listed. This project may therefore continue for some time. In fact, if new information becomes available to the

Comparing Fractions, Decimals, and Percents

students, they should be allowed to add it to existing vertical lines or to insert additional lines. For this reason, it is suggested that the resulting Fraction/Decimal/Percent/Applications Chart (see figure below) (plus some blank space for additions) be taped to the wall to allow for continued work. (*Note:* These charts have often been placed above chalkboards or bulletin boards, and the students have been allowed to use a step ladder to add items and record new findings.)

Example:

These students below are working cooperatively to mark in portions of their Fraction/Decimal/Percent/Applications Chart. Comments, like those the students have made below, are often very helpful in determining learners' "true" levels of understanding.

Chapter 24

Number Clues

Grades 4–8

☒ Total group activity
☒ Cooperative activity
☒ Independent activity
☐ Concrete/manipulative activity
☐ Visual/pictorial activity
☒ Abstract procedure

Why Do It:

Number Clues helps develop students' number sense by emphasizing the relationships between numbers, and enhances their comprehension of mathematical terms.

You Will Need:

One index card for each clue, one index card for each individual number, and one index card as a scorecard are required. One sample "Number Clue Activity" that can be duplicated, cut out, and tried is provided. Samples of other "Number Clue" activities are also provided in the Extensions, and can be placed on index cards.

How To Do It:

1. It is best to do the activity with groups of three or four players, but it can be done with the whole class or even with one individual player. The purpose of the activity is to eliminate numbers as the clues are read, and to ultimately find the one number that satisfies all clues.

2. The clue cards are passed to each individual player in a group. If there are four clue cards and only three players, one player will receive two clues. The number

cards are placed face up in the middle of the group. The scorecard is numbered 1 to 4, and used to keep track of each answer for the four different games. The player with Clue #1 reads his or her card out loud, and then uses the information on the card to take away any numbers from the center that do not satisfy the clue. The player with Clue #2 then reads his or her clue card and uses this clue to take away another number or numbers from the middle. The game continues until there is only one number left in the middle and all clues have been read.

3. The group should double-check to see that the number left in the middle satisfies all the clues. The group will then record their answer on the scorecard.

4. Distribute a new set of cards to the group to start another game. There are usually four games for each activity.

5. After finishing the entire activity (four games in all), the group will receive a point for every correct answer on their scorecard. If other groups are playing at the same time, scores can be placed on the chalkboard. If time permits, or on another day, the same groups could play again and scores could be totaled. The group below has already eliminated 7 and 25, and Fay is reading her clue.

Example:

Provided at the end of this chapter is a complete "Number Clue Activity" consisting of four games, complete with cutout numbers and clues that can be photocopied. The answers to this set of four games are: Game 1, 24; Game 2, 81; Game 3, 89; Game 4, 15.

Extensions:

1. Games can be developed using fractions, decimals, percents, integers, and other algebraic concepts. Some numbers and sample clues are provided below. Answers to samples are: Sample 1, 135; Sample 2, 36; Sample 3, 2/3; Sample 4, 3/5; Sample 5, 5/6; Sample 6, 0.425. (*Hint:* When doing Samples 3 and 4, changing fractions to have a common denominator works well. For example, in Sample 3, the least common denominator of 120 works well, and in Sample 4 finding different common denominators along the way, as numbers are eliminated, is preferred. Also, when doing Samples 5 and 6, changing all numbers to decimal form is a common method.)

2. Students can be challenged to make up their own clues for a set of numbers.

Sample Number Clue Games for Whole Numbers:

Sample 1

Number Possibilities: 54   60   135   75   180

Number Clues:

Clue #1:   It has 5 as a factor.

Clue #2:   It is a multiple of 3.

Clue #3:   It is <u>not</u> a multiple of 10.

Clue #4:   The sum of the digits is 9.

Sample 2

Number Possibilities: 36   72   216   716   63

Number Clues:

Clue #1:   It has 4 as a factor.

Clue #2:   It is a multiple of 9.

Clue #3:   The product of the digits is greater than 12.

Clue #4:   It has exactly nine factors.

Sample Number Clue Games for Fractions:

Sample 3

Number Possibilities: 2/3   1/2   5/8   4/5   3/4

Number Clues:

  Clue #1: It is > 3/5.

  Clue #2: It is < 23/30.

  Clue #3: The denominator is one more than the numerator.

  Clue #4: The denominator is a prime number.

Sample 4

Number Possibilities: 3/6 2/3 6/9 1/2 3/5 1/6

Number Clues:

  Clue #1: It is reduced to its lowest terms.

  Clue #2: It is between 9/20 and 4/5.

  Clue #3: It is > 4/7.

  Clue #4: It is < 11/18.

Sample Number Clue Games for Fractions, Decimals, and Percents:

Sample 5

Number Possibilities: 5/6 7/10 1/2 3/2 2/3 11/12

Number Clues:

  Clue #1: It is > 0.7.

  Clue #2: It is < 0.999 . . .

  Clue #3: Its decimal equivalent repeats.

  Clue #4: It is less than 85%.

Sample 6

Number Possibilities: 33 1/3% 25% 50% 0.425 0.1666

Number Clues:

  Clue #1: It is >1/5.

  Clue #2: It is <1/2.

  Clue #3: The decimal form of the number terminates.

  Clue #4: The digit in the hundredths place is less than 5.

Number Clue Activity

GAME 1 NUMBER CARDS

7	15	36	25	24

GAME 1 CLUE CARDS

Clue #1 It is a two-digit number.	Clue #2 It is a multiple of 3.
Clue #3 The sum of the digits is less than 9.	Clue #4 The ones digit is twice the tens digit.

GAME 2 NUMBER CARDS

9	81	36	25	27

GAME 2 CLUE CARDS

Clue #1 It is an odd number.	Clue #2 It is a two-digit number.
Clue #3 It is a perfect square number.	Clue #4 It is divisible by 9.

GAME 3 NUMBER CARDS

41	59	67	87	89

GAME 3 CLUE CARDS

Clue #1	Clue #2
It is a prime number.	The ones digit is greater than the tens digit.
Clue #3	**Clue #4**
The sum of the digits is a prime number.	One of the digits is a perfect square.

GAME 4 NUMBER CARDS

7	15	27	35	45

GAME 4 CLUE CARDS

Clue #1 It is a two-digit number.	**Clue #2** It is a multiple of 3.
Clue #3 It has a factor of 5.	**Clue #4** The product of the digits is less than 16.

Chapter 25

Number Power Walks

Grades 4–8

☒ Total group activity

☒ Cooperative activity

☐ Independent activity

☒ Concrete/manipulative activity

☒ Visual/pictorial activity

☒ Abstract procedure

Why Do It:

Students will physically act out and conceptualize the powers of numbers.

You Will Need:

No equipment is required, unless precise measurements are desired. Measuring devices, such as yardsticks or meter sticks, long tapes, or trundle wheels, in addition to chalk, can be used.

How To Do It:

1. Be certain the players understand that a power of a number is the product obtained by multiplying the number by itself a given number of times. For example, to square the number 3 (also called raising 3 to the second power), means to treat it as 3^2 or 3×3, yielding 9. Likewise, 3^3 (read as 3 to the third power or 3 cubed) yields $3 \times 3 \times 3 = 27$. As soon as the players have a basic grasp of these mathematical ideas, they are ready to act them out.

2. Have the players stand in groups of four behind a starting line. Note that for the first round they will ''walk off'' number power distances for the number 2:

the first participant from each group will walk forward 2^1 paces, the next individual 2^2, the third person 2^3, and the fourth group member 2^4; the individuals will have walked forward 2, 4, 8, and 16 steps, respectively. Then ask, "How far would someone going 2^5 steps need to travel?" When the players agree on an answer, select someone to walk it off. Then continue, asking, for example, about 2^6 or 2^7.

3. The number of necessary steps will eventually become too great to walk off in a straight line if students are to remain on the school grounds. At this point have the players discuss and agree on an estimate of where several more powers for that number would place an individual. Next, try another number, perhaps 3 or 4, this time "hopping off" the number power distances. Vary the physical activity for each new Number Power Walk and, if greater precision is desired, make use of trundle wheels, long tapes, or other measurement tools. After completing several such walks, the players not only will have gained a firm understanding of the powers of numbers but also will have enjoyed the experience.

Example:

In the illustration shown, the players have made Number Power Walks of $2^1, 2^2, 2^3,$ and 2^4 paces.

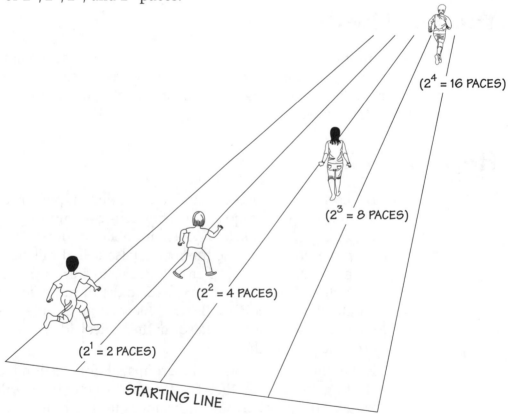

(2^4 = 16 PACES)

(2^3 = 8 PACES)

(2^2 = 4 PACES)

(2^1 = 2 PACES)

STARTING LINE

Extensions:

1. Try a situation in which the powers remain constant but the base numbers sequentially increase in size. For example, have students determine what will result when a series of numbers is cubed, such as $2^3, 3^3, 4^3, 5^3$, and so on.

2. When working with such large numbers as 10^2 and 10^3 or 50^3 and 50^4, it quickly becomes impractical to try to act out the results. In such cases, have students mentally estimate the number power distances and discuss where they might end up if they actually took Number Power Walks.

Computation Connections

The activities in this section will help students understand more than just how to perform the operations of addition, subtraction, multiplication, and division. Your students will have direct hands-on and visual experiences that will enable them to explore how and why computation procedures work. As a result, they will develop conceptual understanding as they practice computation through these interesting, informative, and fun tasks.

Selected activities from other portions of this book can also be used to help reinforce learners' computation understandings. Some of these are *Beans and Beansticks* (p. 13), *Post-it Mental Math* (p. 47), and *Reject a Digit* (p. 57) in Section One; *Verbal Problems* (p. 260) and *Student-Devised Word Problems* (p. 274) in Section Three; and *Magic Triangle Logic* (p. 358) and *Dartboard Logic* (p. 397) in Section Four.

Chapter 26

Paper Clip Addition Cards

Grades K–2

☐ Total group activity
☒ Cooperative activity
☒ Independent activity
☒ Concrete/manipulative activity
☒ Visual/pictorial activity
☒ Abstract procedure

Why Do It:

This activity gives students experience with visual and hands-on devices that help in the understanding of the abstract addition and subtraction number facts.

You Will Need:

Boxes of paper clips (possibly one box per child or group of children), large index cards, and markers are required.

How To Do It:

In this activity, students will be clipping paper clips on index cards to make a visual representation of an addition or subtraction fact. For example, if a student is given 4 paper clips, he or she could clip 3 on the top left side of the card and 1 on the top right side of the card. This would show that $3 + 1 = 4$ (see figure).

Have students use paper clips and index cards to solve basic addition problems and keep a record of the combinations found on separate pieces of paper. Have them try the problems in the Extensions, and also make up some problems of their own to share with other players.

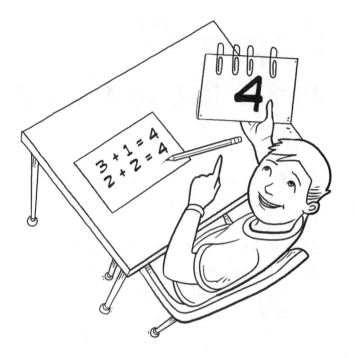

Example:

The combinations for the number 5 are shown with paper clips below. Notice that the paper clips show $4 + 1 = 5$ and $1 + 4 = 5$ are different. This reinforces the commutative property of addition that says $a + b = b + a$ for any real numbers a and b.

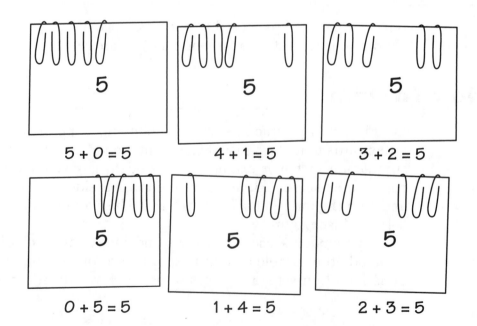

Extensions:

Ask students to use paper clips on number cards to solve the following problems; have them show their work and record answers in number sentences.

1. Use paper clips to find all of the combinations for 4.

2. Show and record the combinations for 7, 10, and 14.

3. Sometimes you will need to record number combinations in 10s and 1s. For example, for the number 23, students could place two paper clips on the left and three paper clips on the right, representing two 10s and three 1s. You could also use different color paper clips to distinguish between the 10s and 1s.

4. Explain how these paper clip number cards also show the related subtraction combinations.

Chapter 27

Arm-Lock Computation

Grades K–4

☒ Total group activity

☒ Cooperative activity

☐ Independent activity

☒ Concrete/manipulative activity

☒ Visual/pictorial activity

☒ Abstract procedure

Why Do It:

Students will gain a greater understanding of mathematical computations by physically acting out solutions.

You Will Need:

All that is needed is space in front of a chalkboard or whiteboard for students to stand in groups, and chalk or whiteboard pens.

How To Do It:

In this activity, groups of students will work together to physically display the setup of a basic computation, such as 3 + 2. The display will help the students solve the problem. "Arm-lock" computations are acted out by groups of different sizes; in some cases, the solutions will involve up to the total number of students in a class (approximately 30). Review the solutions for the examples that follow, and then have students solve similar problems. Finally, if they are ready, students might solve more complex problems (those requiring more than 30 students) by following procedures outlined in the Extensions section.

Examples:

1. In the situation shown below, students were asked to "lock arms" with 3 in the first group and then with 2 in another group. Next they were asked to figure out the total number of players with locked arms by "adding" the 3 students and the 2 students. (If they wished, they might also have "arm-locked" the two groups together to show that when the two numbers are added, you can combine the students into one group to find the solution.) Notice that the related equations for $3 + 2$ were written on the chalkboard.

2. The following group of students have been posed with the problem $9 \div 4$. To do the Arm-Lock Computation, the 9 students get into groups of 4. The students discovered that they could form 2 groups of 4, with 1 student left over, and therefore by "dividing" 9 by 4, this led to 2 groups with a remainder of 1. This could lead to the discussion that when we divide 9 by 4, we are looking for how many groups of 4 are in the number 9. Furthermore, the written computations for $9 \div 4$ have been recorded for everyone to see on the chalkboard. (*Note:* This example denotes measurement division, where students make 4 the number in each set and then count the sets. For comparison, the students might also be asked

Arm-Lock Computation

to use the same numbers in a partitive division problem, in which the students would form 4 sets with the same number in each set. This method would yield 4 groups of 2 students in each, plus a remainder of 1 student.)

Extensions:

Try some of the following procedures as a class. Then have students create problems of their own or solve some from their math workbooks or textbooks.

1. For subtraction problems, begin with the total number of students in arm-lock position and ask those to be "subtracted" to sit down or move out of sight. The remaining students will represent the correct answer. (*Note:* This situation calls for take-away subtraction; the students might also use the same numbers in a comparative subtraction situation, in which the groups line up side by side and compare to see "what the difference is.")

2. Multiplication problems will require students to get into groups with the same number of students in each. For $3 \times 4 =$ ____, for example, they should organize as 3 groups with 4 students' arms locked in each group.

3. Computation with larger numbers can be accomplished by bringing several classes together. In order to solve a problem such as 56 divided by 7, first have 56 students line up and lock arms. Then have them count players up to 7, with the 7th person unlocking from his or her neighbor (resulting in 7 students' breaking away from the larger group). Continue and count to 7 again from the remaining group, with the 7th person also unlocking from his or her neighbor. Repeat the process until every student has been counted, at which time the number of arm-locked groups with 7 students each are counted to show that 8 groups of 7 yield 56 students.

4. For students who are ready to move beyond 1-to-1 correspondence, problems can be solved with the help of place value cards reading 10s; 100s; 1,000s; and so on. With a problem like 123 × 4 = ____, for example, 4 groups each would have 1 student holding a 100s card, 2 students with 10s cards, and 3 individual students holding no cards. The groups could then be combined because multiplication is repeated addition, namely 123 + 123 + 123 + 123. The combined groups would total 4 students with 100s cards, 8 with 10s cards, and 12 single players, or 400 + 80 + 12. Then, after "trading" 10 of the individual players for 1 with a 10s card, the product is shown to be 400 + 90 + 2 = 492.

Chapter 28

Punchy Math

Grades K–6

☒ Total group activity
☒ Cooperative activity
☒ Independent activity
☒ Concrete/manipulative activity
☒ Visual/pictorial activity
☒ Abstract procedure

Why Do It:

Students will have the opportunity to concretely "prove" the results of number computations.

You Will Need:

A paper hole punch, scrap paper, and pencils or crayons are required for each group of students.

How To Do It:

For this activity, students will be punching holes and looping sets of holes to perform a basic computation. Looping a set involves drawing a "circle" around a group of punched holes in order to consider them as one set. This procedure will allow students to see the numbers in an abstract problem as sets of holes and can then add the sets together to find the answer to the problem.

Examples:

1. For the problem $3 + 2$, students were instructed to punch 3 holes along one edge of the paper and use their pencils to loop this group of holes and label with

the number 3. Then they punched, looped, and labeled 2 more holes next to the 3 already punched. After doing so, students were directed to draw a large loop around all of the punched holes and count the total. Finally, the mathematical sentence was written as shown.

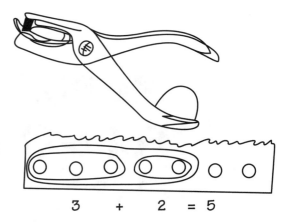

3   +   2   =   5

2. For the problem 3 × 7, students folded sheets of paper into 3 layers and punched 7 holes. The paper was opened, looped, and labeled to show 7 + 7 + 7 or 3 × 7 = 21. By turning the same punched paper sideways and drawing loops in the other direction, it is also possible to show 3 + 3 + 3 + 3 + 3 + 3 + 3 or 7 × 3 = 21.

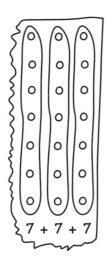

7 + 7 + 7

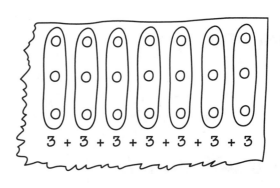

3 + 3 + 3 + 3 + 3 + 3 + 3

Extensions:

Have students punch, loop, and label the following problems. Then have them explain how they did each problem and how they can be certain of the correct outcome.

1. 3 + 5	2. 4 + 7	3. 2 + 2 + 2 + 2	4. 4 × 2	5. 2 × 4
6. 4 × 6	7. 5 × 8	8. 6 × 6	9. 4 × 9	10. 3 × 12

After they have completed these problems, have students choose any other problem they wish and then punch, loop, and label it, and "prove" that their outcome is correct.

Chapter 29

Multiplication Fact Fold-Outs

Grades K–6

☐ Total group activity
☒ Cooperative activity
☒ Independent activity
☐ Concrete/manipulative activity
☒ Visual/pictorial activity
☒ Abstract procedure

Why Do It:

Students will construct and use visual paper fold-outs that will help them conceptualize and reinforce multiplication facts using repeated addition.

You Will Need:

This activity requires several lightweight tagboard strips, approximately 3 by 24 inches (may be cut from 18- by 24-inch stock) for each participant; marking pens; and a large supply of identical stickers (optional).

How To Do It:

1. To construct a Fact Fold-Out, begin by folding and creasing a lightweight tagboard strip. The first fold should equal 1 inch; each successive fold is a little larger. Make the next fold over the top of the first (for example, like a toilet paper roll) and continue in this manner until the entire strip is folded and creased.

2. Next, use a marking pen to both write sequential fact problems and draw the associated visual images (or use

stickers for the visuals). Do so by unrolling one segment of the fold-out and writing 1 × *(factor being studied)* = _____ on the "fat portion or unfolded part" and drawing the related visual image on the "flap." Unfold a second segment, and on the new fat portion write 2 × *(factor being studied)* = _____, drawing a repeat image (same image as drawn on the first flap) on the newly exposed flap segment. Continue in this manner until reaching 10 or 12 × *(factor being studied)* = _____. After cutting off any excess tagboard, the fold-out will be ready for use.

3. After you have constructed a fold-out and demonstrated how to do so, each student should, over a period of time, construct several of his or her own. Students should utilize the strips to practice the multiplication facts on their own, as part of a group, and with parents or other adults. Note that they are not only practicing facts but also internalizing a visual image of "how many" each time they make use of the fold-outs.

Examples:

1. The fold-out below was constructed to give students both visual and abstract practice with multiplication facts for the 3s. The fold-out, including both frog stickers and numbers, is unrolled first to show 1 × 3 = 3, then 2 × 3 = 6, and further 3 × 3 = 9.

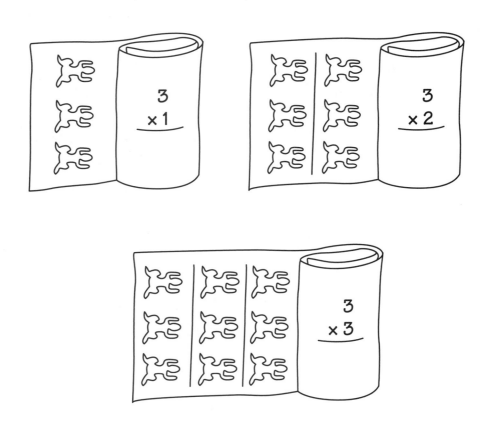

2. The following fold-out is demonstrating $6 \times 5 = 30$.

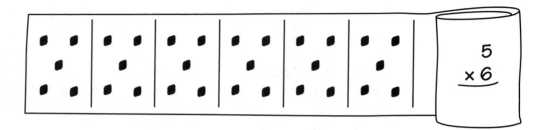

Extensions:

1. Young students might make use of similar fold-outs to study repeated addition. The 2×3 in Example 1 above, for example, might be best understood as $3 + 3$, and the 3×3 taken as $3 + 3 + 3$.

2. Use calculators in conjunction with the fold-outs. For example, have the players press $3 + 3 =$ ____, $=$ ____, $=$ ____, $=$ ____, $=$ ____, and so on, and keep a record of the displayed answers. Discuss how the calculator answers compare to those for multiplication or repeated addition on the fold-outs.

3. Make Multiplication Fold-Outs for any facts (up to 10×10 or 12×12) that need practice. Demonstrate to students how these fold-outs can be used to check whether or not an answer is correct.

Chapter 30

Ziploc™ Division

Grades K–6
- ☐ Total group activity
- ☒ Cooperative activity
- ☒ Independent activity
- ☒ Concrete/manipulative activity
- ☒ Visual/pictorial activity
- ☒ Abstract procedure

Why Do It:

Students' comprehension of division will be enhanced through manipulating objects, visual displays, and abstract connections.

You Will Need:

This activity requires several large Ziploc bags, marbles, large beans, or centimeter cubes of the same color (100 or more). Also required are masking tape and one marking pen per group.

How To Do It:

Place designated numbers of marbles (for example, 10 or fewer for beginners and up to 100 for advanced students) in several Ziploc bags, squeeze the excess air out, and seal the bags. Using the marking pen on the tape, write separate division instructions for each marble set and tape them to the bags. The students should then arrange the marbles into sets of the size called for in the instructions by pushing them around without opening the bags. When finished manipulating, the students should record their findings either as sentences or division algorithms and be ready to explain their outcomes. For example, when computing $12 \div 4$, the students could divvy the marbles into 4 different groups with 3 in each group. Or, the students could push 4 marbles to one corner of the bag, 4 marbles to another corner, and so on, resulting with 3 groups of 4 marbles. Finally, the equation $12 \div 4 = 3$ is written on the tape and taped to the bag. For further clarification, see the Examples below.

Examples:

1. The Ziploc bag shown below contains 10 marbles. The instructions, as written on the masking tape, say, "Arrange the marbles in sets of 2. Then write a sentence about what happened."

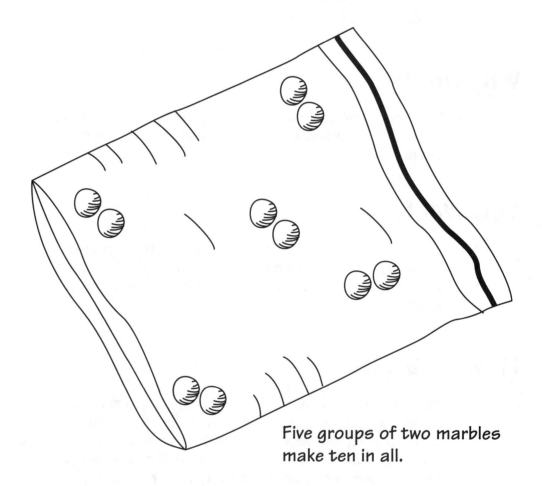

Five groups of two marbles make ten in all.

2. The next bag contains 45 marbles. The directions on the masking tape state, "Place these marbles in groups of 6. Write your answer as a division problem. Then explain your work to another student."

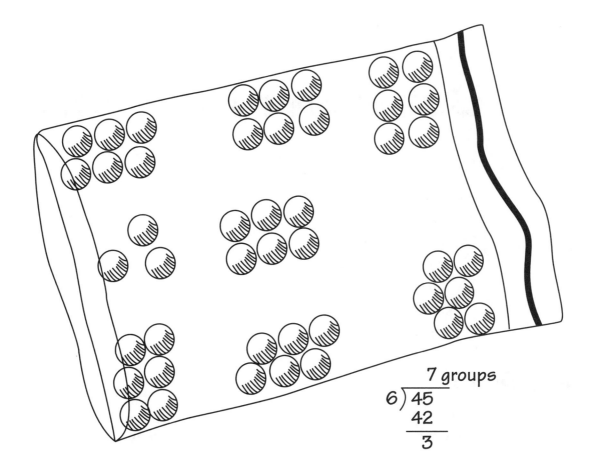

7 groups
6)45
42
3

Extensions:

1. Beginning students can approach this activity in an informal manner, simply manipulating the marbles within the bags and discussing into how many groups of a certain size they were able to divide the entire set of marbles. If they are ready, they can also keep records of their findings by writing simple word sentences or equations.

2. The Examples above both involve measurement division (you know the number in a set, but not the number of sets). Students should also experience partitive division (you know the number of sets, but not the number in each set). Reworked as a partitive problem, Example 1 above, with 10 marbles, would instead have the directions, "Arrange the marbles into 5 groups of the same size. How many marbles are there in each group?"

3. Advanced players might complete problems involving large numbers of marbles. A partitive division situation, for example, might require 234 marbles to be divided into 7 groups. The 234 marbles would be distributed 1 at a time to each of 7 bags, with the outcome showing 33 marbles per bag and 3 extras.

Chapter 31

Dot Paper Diagrams

Grades K–6

☒ Total group activity
☒ Cooperative activity
☒ Independent activity
☐ Concrete/manipulative activity
☒ Visual/pictorial activity
☒ Abstract procedure

Why Do It:

Students will bolster their understanding of computation by bridging visual representations with abstract procedures.

You Will Need:

This activity requires dot paper (reproducibles with 50, 100, over 1,000, and 10,000 dots are provided), colored markers, and a pencil.

How To Do It:

In this activity, students use dot paper to show the solutions for addition, subtraction, multiplication, or division problems. If a student is asked to solve the problem $20 - 4 =$ ____, then he or she would start by looping 20 of the dots on a piece of dot paper. To subtract 4 from this set of 20 dots, the student would then cross out 4 dots within the set using a pencil. The number of dots in the looped set of 20 that are not crossed out is the answer to the problem. After giving students the Examples here, ask them to attempt the problems in the Extensions section by looping or marking the number values on the dot paper. Sometimes it is helpful for students to use markers of several different colors. Also, have them show their numerical computations and answer for each problem.

Examples:

1. Students might show the problem $4 + 5$ by looping 4 dots in one color and 5 adjacent dots in another. They then should draw a larger loop around both to reveal $4 + 5 = 9$.

2. A grid with more dots is required to figure out $27 - 9$. Students must first loop 27 dots, and then cross out 9 of these. The remaining dots show the answer.

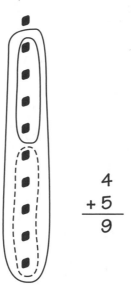

$$\begin{array}{r} 4 \\ + 5 \\ \hline 9 \end{array}$$

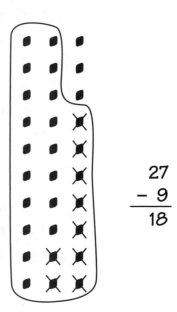

$$\begin{array}{r} 27 \\ - 9 \\ \hline 18 \end{array}$$

3. The multiplication problem 5×9 can be shown in more than one way. An efficient method (that avoids having to recount the dots) is to use every column of ten dots, as shown in the figure.

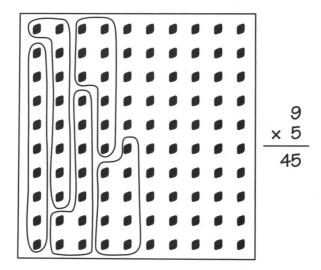

$$\begin{array}{r} 9 \\ \times 5 \\ \hline 45 \end{array}$$

4. Such a division problem as $197 \div 30$ will require two 100-dot areas. Students first loop 197 dots and then subdivide these into

Dot Paper Diagrams

groups of 30. The number of full groups of 30 is the whole number answer, and the area with fewer than 30 dots is the remainder.

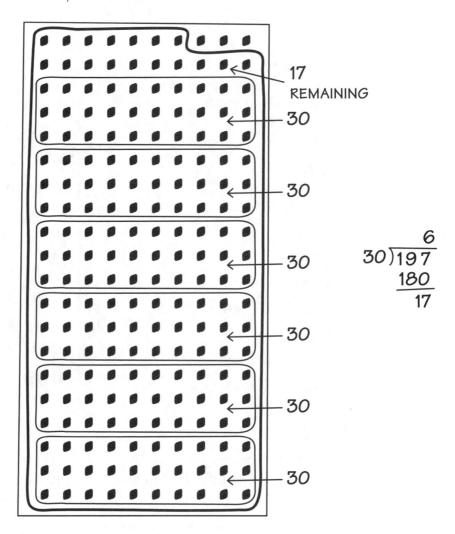

$$30\overline{)197} \\ \underline{180} \\ 17$$

with quotient 6 and remainder 17.

Extensions:

Have students use dot paper to solve as many of these problems as they can, requiring them to show their numerical computations and answers for each problem.

1. $7 + 2 =$

2. $\begin{array}{r} 27 \\ +14 \\ \hline \end{array}$

3. $\begin{array}{r} 9 \\ -3 \\ \hline \end{array}$

4. $88 - 45 =$

5. $6 \times 7 =$

6. $\begin{array}{r} 23 \\ \times 4 \\ \hline \end{array}$

7. $\begin{array}{r} 179 \\ -87 \\ \hline \end{array}$

8. $9\overline{)27}$

9. $155 \div 20 =$

10. Create three or four of your own problems. Share them with a friend.

50 DOT PAPER

100s Dot Paper

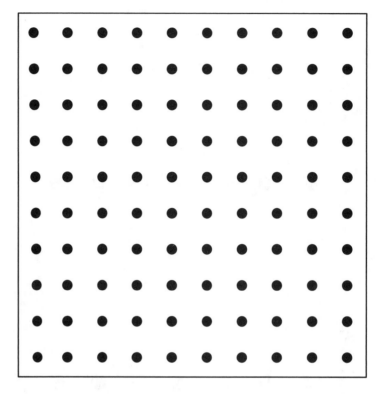

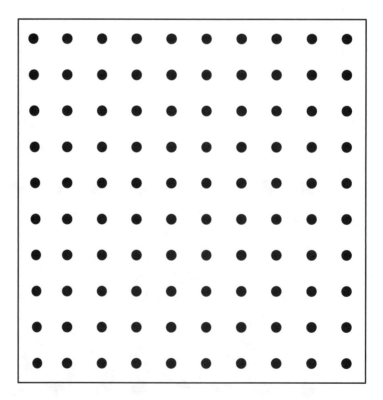

Dot Paper for 1,000 and More

Dot Paper Diagrams

10,000 Dots

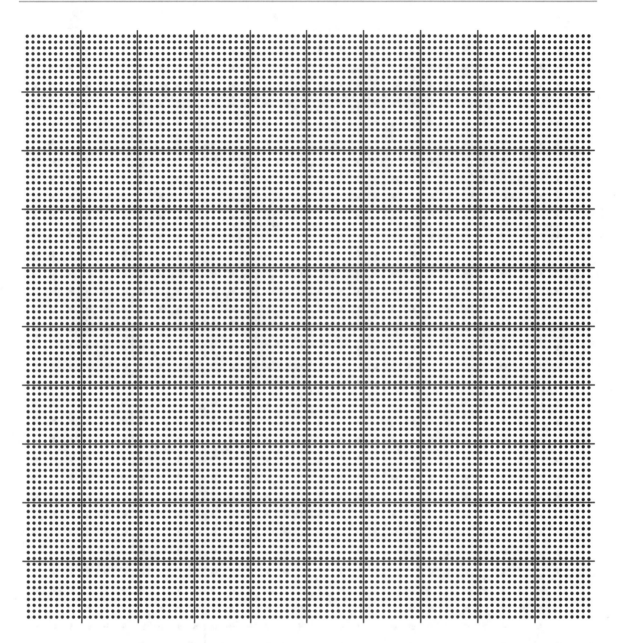

Chapter 32

File Folder Activities

Grades K–6
☐ Total group activity
☒ Cooperative activity
☒ Independent activity
☐ Concrete/manipulative activity
☐ Visual/pictorial activity
☒ Abstract activity

Why Do It:

This activity gives students practice with basic facts, computation, and problem solving by matching. It encourages students to work as individuals or in small groups and to self-check their answers. It can also be used as an alternate way for the teacher to test the students.

You Will Need:

Students will need manila file folders (amount will vary depending on what you feel needs reviewing), glue, library pocket envelopes (pockets made out of construction paper or 4- by 6-inch index cards will also work), 3- by 5-inch index cards, and marking pens.

How To Do It:

1. This activity involves the use of file folders, pockets, and cards to help students review basic mathematical concepts or to test their knowledge in basic mathematical facts.

   Glue the library pockets inside the file folders. The number of pockets depends on what is needed for the review or test. For example, if students need to review their multiplication facts, then their folder could look like the one pictured in Example 1. Use the marking pens to write problems on the pockets and their answers on individual index cards. Prior to starting, hide an answer key to all the pockets in the Answer Cards pocket.

2. To begin, each student (or pair of students) must take all the individual answer cards and place them in the matching problem pockets. If students need to make pencil-and-paper computations for a given problem, they should use a small piece of scratch paper and place it in the same pocket as the answer card. Solutions should be checked against the answer key or by another player (sometimes with a calculator). If you choose to use the file folder as a test of a student's knowledge, then the activity could be timed, with you checking the solutions.

Examples:

1. Using the file folder below, a student must match multiplication facts with the corresponding answers.

2. This file folder requires students to practice telling time from a clock face. Students match digital times to those on "regular" clock faces.

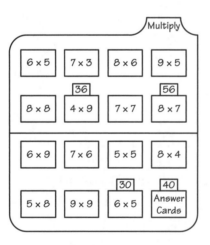
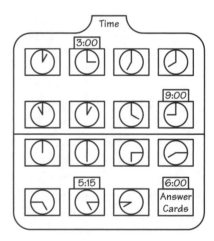

Extensions:

1. Devise file folders for any area in which practice is needed. The players, after seeing how the folders are constructed, should make a variety of folders for one another. These file folders can be made for matching numerals with pictured amounts; practicing basic facts; exploring fractions, decimals, measurement, and geometric identification; working with the concepts of time and money; and solving short story problems, among other things.

2. Older students can construct and use these file folders to provide special help for younger learners. Especially useful are folders dealing with numerals, number sense, place value, and basic math facts. The folders could also be at a math station in the classroom, and as students finish their regular work, they could go to the station to practice certain skills.

3. To increase the difficulty level for this activity and promote careful thinking, include more than a single correct answer card for certain problems (for example, a pocket reading "Find two numbers whose product is 36" might be answered with 4×9 and 6×6) or include a few wrong answers that do not correspond to any of the folder problems. One or both of these tactics can be used to turn this activity into a math quiz.

Chapter 33

Beat the Calculator

Grades 1–6

☒ Total group activity
☒ Cooperative activity
☐ Independent activity
☐ Concrete/manipulative activity
☒ Visual/pictorial activity
☒ Abstract procedure

Why Do It:

Students will practice basic facts and mental math. This activity will help students become more proficient in recalling basic mathematical facts, and with this ability they will have fewer problems with more complicated mathematics.

You Will Need:

At least one calculator is required. If *Beat the Calculator* is to be a large group or whole class activity, the teacher can use a calculator placed on an overhead projection device, or even a virtual calculator found online. For small groups of two to five participants, only one calculator is necessary.

How To Do It:

1. In a small group, with one calculator and three people, the following procedure works well: Student 1 calls out a math problem, such as 6×7. Student 2 uses a calculator to solve the problem and state the answer. At the same time, Student 3 solves it mentally and says the answer. The first to give the correct answer (Student 2 or Student 3) wins. The players' roles should eventually be rotated. To make the activity more competitive, have students tally the number of wins for each student.

(*Note:* Students will soon discover that if they have practiced their basic facts, they will be able to "beat the calculator" nearly every time.)

2. *Beat the Calculator* may also be played as a whole class activity. In this case, you or a leader operates the calculator, while students simultaneously do the mental math and call out answers; a chosen judge calls out the problems and determines the winner of each round. The object of the activity is therefore to determine which method is faster and more efficient for obtaining the solutions to basic fact problems: using a calculator or just memorizing the fact.

Example:

In the small group situation above, Sean had called out 5 × 9. Susan has been attempting to solve the problem with a calculator before Randy could do so mentally. However, because Randy had mastered his 5s multiplication facts, he was able to beat the calculator.

Extensions:

1. Young students might try counting with a calculator by entering a number (try 1), an operation (try +), and pressing the equal button multiple times (= = = =) to make the calculator count by 1s. They can also start with any number, like 20, and enter 20 + 1 = = = =. They might further try counting forward (addition) or backward (subtraction) by any multiple; for example, they can

enter $3 + 3 = = = =$ and see what happens. (*Note:* Learners should use a calculator that has an automatic constant feature built in; most basic calculators do. To test this, simply try the calculator; if it "counts" as noted above, it has the needed constant.)

2. Students might use a calculator that has an automatic constant to individually practice basic multiplication facts. For example, to practice the 4s facts, they can enter $4 \times =$, which the calculator holds in its memory. Then any number entered will be multiplied by 4 when the $=$ key is pressed. Thus the student might enter 8, mentally think what the answer should be, press $=$, and see that the answer displayed is 32.

3. Advanced students can work in pairs, taking turns trying to beat the calculator with such tasks as $2 \times 12 \div 8 + 3 - 5 = \underline{\quad}$ or $7 + 32 \div 4 - 5 \times 2 = \underline{\quad}$. (*Note:* In these cases, be certain that the players use the proper order of operations. The mnemonic "Please Excuse My Dear Aunt Sally" sometimes helps students remember the order: parentheses, exponents, multiply or divide from left to right, and add or subtract from left to right. Students are frequently unclear about this concept, so even if the calculator has a built-in order of operations feature, the students should be taught to put parentheses as shown here: $7 + (32 \div 4) - (5 \times 2)$. The answers to the two problems in this Extension are 1 and 5, respectively.

Chapter 34

Floor Number Line Actions

Grades K–6
- ☐ Total group activity
- ☒ Cooperative activity
- ☐ Independent activity
- ☒ Concrete/manipulative activity
- ☒ Visual/pictorial activity
- ☒ Abstract procedure

Why Do It:

Students will physically act out computation and mathematical problem-solving situations.

You Will Need:

This activity requires a walk-on number line, which can be constructed using soft chalk, tape, and number cards, or a large roll of paper with a marking pen.

How To Do It:

1. To construct the number line, write large numerals about 1 foot apart either on the playground or floor using soft chalk, or on a large roll of paper using a marking pen. If the number line will be used more than once, it can be made by taping number cards to the floor or using some more permanent method on the playground surface. Problems at the beginning will likely make use of the numerals 0 through 10, but as the work becomes more difficult, the number line can be expanded to 100

or more. If signed numbers are to be used, it should also be extended from 0 to $-1, -2, -3$, and so on.

2. Students will solve math problems using this walk-on number line. The examples below show how to begin. Once students understand the procedure, have them try some of the problems in the Extensions section or other problems created by you or the students. Students should be ready both to explain how they "walked out" each problem and to use pencils and paper to show the same solution.

Examples:

1. For $4 + 3$, students begin at 0 and take 4 steps to the number 4. Then they take 3 more forward steps and check the number on which they are now standing; it should be 7. (See the solid arrow in the illustration below.) Finally, students should keep a record by writing $4 + 3 = 7$ in their notebooks.

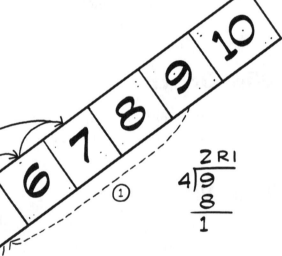

2. For such a problem as $9 \div 4$, students begin at the 9 and move toward the 0, taking 4 steps at a time and holding up a finger for each time. Beginning at the 9, they step to 8, 7, 6, and 5 and hold up 1 finger; then they step to 4, 3, 2, and 1 and hold up 2 fingers. They have therefore taken 4 steps 2 times, but still need to get to 0; this will require 1 more step. Thus $9 \div 4$ requires 2 sets of 4 steps with 1 step remaining, so $9 \div 4 = 2$, with a remainder of 1.

Computation Connections

Extensions:

Have students use the floor number line to help solve either the following problems or others that students need to solve from a workbook or text. Students can also make up several of their own problems and have their classmates use the number line to figure out the solutions.

1. $8 + 3$	2. $7 - 4$	3. 4×6
4. $20 \div 5$	*5. $-2 + -4$	*6. $-3 + 4$

*(*Hint:* Face in the direction of the first signed number, and then change direction every time the sign of the number changes.)

Chapter 35

Egg Carton Math

Grades K–6

⊠ Total group activity

⊠ Cooperative activity

⊠ Independent activity

⊠ Concrete/manipulative activity

⊠ Visual/pictorial activity

⊠ Abstract procedure

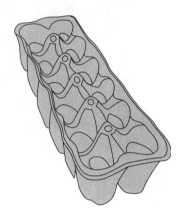

Why Do It:

Egg Carton Math provides students with a fun way to generate problems and practice computation, and helps them gain a better understanding of factors of a number, prime factorization, and probability.

You Will Need:

A 12-cell or 18-cell egg carton with its lid still attached is required for every student or pair of students. Also required are paint and brushes, card stock, glue or tape, sticky dots for labeling, crayons or markers, pencils, beans (larger beans preferably, such as pinto or lima beans), and photocopies of the "Egg Carton Probability" handouts (provided).

How To Do It:

There are many things you can do with an egg carton; here are a few activities designed for various levels of math ability.

1. In order to learn about probabilities, students will be shaking an egg carton with a bean inside. The cells of the egg carton will be painted different colors, and the probability of the bean landing on a certain color is discussed.

Begin with the students painting the egg carton cells with three different colors in random order. Either tell students how many cells will be painted red, how many green, and how many blue, or let them decide for themselves. The colors might also be chosen randomly by tossing a die: if 1 or 4 appears, students paint the cell red; if 2 or 5 comes up, they paint the cell green; and if 3 or 6 is rolled, they paint the cell blue. With this method, it is possible that students will never paint a cell red, which simply means that the probability of getting a red is zero. Students then need to cut a piece of card stock the size of the top of the egg carton and glue it to the top to block the holes through which beans might slip. They then place one bean inside the egg carton and close the lid. Students will use the "Egg Carton Probability" activity sheet to perform the probability experiment. After recording their results, students can also draw a bar graph on the back of their activity sheets. Younger students can fill in sections of the blank bar graph included at the end of this activity.

2. Another way to use the egg carton is to write the numbers 1 through 12 or 1 through 18 (depending on the size of the egg carton) on sticky dots, and put the dots in each cell such that the numbers show. Students can use two beans, three beans, or more depending on their math ability level. After the student shakes the egg carton, he or she opens it and writes down the numbers on which the beans have fallen. If two beans land in the same cell, the student is to write down that number twice. At this point, the student can add or multiply the numbers together. It works best if the students are in pairs, so that they can check each other's answers. If there are just two beans in the egg carton, the activity can help students review basic addition and multiplication facts. For example, one student, after shaking the egg carton, could open it to find the beans have landed on 5 and 6. The student would then ask his or her partner to find the sum of 5 and 6. Students can also keep track of how many correct answers each player comes up with in 10 or 20 shakes.

3. The third activity is for more advanced students and helps enhance students' understanding of prime numbers (counting numbers with exactly two different factors), composite numbers (counting numbers with more than two different factors), and prime factorization. Students begin by writing the first twelve or eighteen prime numbers on the sticky dots (2, 3, 5, 7, 11, 13, 17, 19, 23, 29, 31, 37, 41, 43, 47, 53, 59, and 61). Then they place the dots in individual cells of the egg carton. Students can use two or more beans, shake the egg carton, and use the prime numbers on which the beans fall as prime factors of a composite number. Students can then find the composite number by multiplying the

prime numbers together. They can also practice using exponential notation. For example, a student's five beans could land on 2, 2, 7, 5, and 13. The prime factorization is written $2 \times 2 \times 5 \times 7 \times 13$ or $2^2 \times 5 \times 7 \times 13$ which is equal to the composite number 1,820.

Once students have the prime factorization of a number, they can find all the factors of that number. For example, using the prime factorization of the number 36 (which is $2^2 \times 3^2$), a table as shown below can be set up to find all the factors of 36. Because the exponent of 2 is 2, 0 through 2 will be used for the exponents of 2. Similarly, because the exponent of 3 is 2, the exponents of 3 will range from 0 to 2. Also remind students that any number with an exponent of 0 is equal to 1. To fill out each square in the chart, students will fill in the power of 2 using the exponent in the same row at the left and the power of 3 using the exponent in the same column at the top.

		Exponents of 3		
		0	1	2
Exponents of 2	0	$2^0 \cdot 3^0 = 1$	$2^0 \cdot 3^1 = 3$	$2^0 \cdot 3^2 = 9$
	1	$2^1 \cdot 3^0 = 2$	$2^1 \cdot 3^1 = 6$	$2^1 \cdot 3^2 = 18$
	2	$2^2 \cdot 3^0 = 4$	$2^2 \cdot 3^1 = 12$	$2^2 \cdot 3^2 = 36$

Therefore, the factors of 36 are 1, 2, 3, 4, 6, 9, 12, 18, and 36.

Extensions:

1. Whenever a game requires a random number generator, the egg carton with numbers in it can be used, such as when doing the *Sticky Gooey Cereal Probability* simulation with 12 prizes instead of 6 (see p. 214).

2. Place fractions in the cells and have students practice adding or multiplying fractions. Also, students can discuss the fractional part of the whole that the colors red, green, or blue represent. If each student's egg carton is different, students will be able to see the many fractions that develop from coloring an egg carton. To extend this idea further, students could make a list of all the fractions that are possible when egg cartons are colored in different ways.

3. For advanced students, put positive and negative numbers on the sticky dots in the egg carton and have them add, subtract, and multiply the signed numbers. Algebraic expressions can also be placed in the cells of the egg carton, such as $-5x$, $2x^2$, $8x$, and $-4x^2$, and students can be asked to add, subtract, multiply, or divide the expressions.

Egg Carton Probability

Hypothesis:

Open the egg carton and look at the colors inside. If you closed the egg carton with the bean inside and gave the carton some shakes, on which color do you think the bean would land?

If you were to do this experiment many times, on which color do you think the bean would land the most?

Why do you think this?

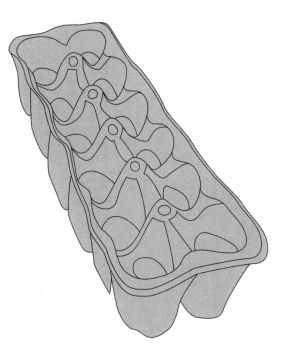

Experiment:

Shake the egg carton with the bean in it. Now open the egg carton and look at what color the bean has landed on. Record the color in the chart below using an X. Do this 20 times, recording your results each time.

Red	
Blue	
Green	

Conclusion:

1. Which color or colors did the bean land on the most, according to your experiment?
2. Did you guess right, when you stated your hypothesis above? Yes or No
3. Draw a bar graph to record your results.

Bar Graph for Egg Carton Probability

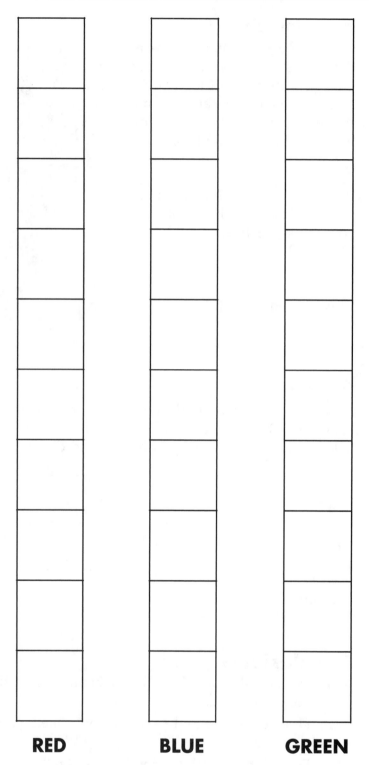

RED   **BLUE**   **GREEN**

Computation Connections

Chapter 36

Cross-Line Multiplication

Grades 2–6

☒ Total group activity

☒ Cooperative activity

☒ Independent activity

☐ Concrete/manipulative activity

☒ Visual/pictorial activity

☒ Abstract procedure

Why Do It:

This activity will visually enhance students' multiplication understandings.

You Will Need:

Students will need pencils and paper.

How To Do It:

This activity requires that students use a pencil to draw crossing lines that correspond to factors (numbers) in any given multiplication problem. If a student wants to solve 3×7, he or she will draw three parallel horizontal lines and seven parallel vertical lines crossing the horizontal lines. Then they will count the number of intersections (line crossings) to find the answer to that specific problem. In this case, the number of intersections is 21 and that is the answer to 3×7. See the figure for clarification.

Example:

To solve the problem 6×2, the student will need to show 6 groups of 2. After drawing 6 horizontal lines and crossing them with 2 vertical lines, the student will then count 12 crossings, which is the answer to the problem.

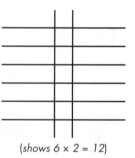

(shows 6 × 2 = 12)

Also, by turning the drawing sideways, the problem 2×6 or 2 groups of 6 can also be shown. Thus, $2 \times 6 = 12$.

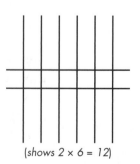

(shows 2 × 6 = 12)

Extensions:

Students could solve the sample problems below.

1. Draw crossing lines to show 3 × 5. There are ____ line crossings. So, 3 × 5 = ____.

2. Show 4 × 5. 4 × 5 = ____.

3. Draw lines for 7 × 5. How many line crossings are there? So, 7 × 5 = ____.

4. Choose a multiplication problem. Draw line crossings for it. Your problem is: ____ × ____ = ____.

5. Figure out some more cross-line problems and share them with the class.

6. Try cross-line multiplication for two-digit multiplication.

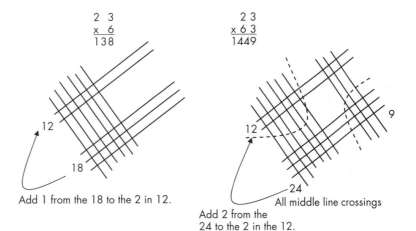

$$\begin{array}{r} 2\ 3 \\ \times\ 6 \\ \hline 138 \end{array}$$

12

18

Add 1 from the 18 to the 2 in 12.

$$\begin{array}{r} 2\ 3 \\ \times 6\ 3 \\ \hline 1449 \end{array}$$

12

9

24

All middle line crossings

Add 2 from the 24 to the 2 in the 12.

Chapter 37

Highlighting Multiplication

Grades 2–6

☒ Total group activity
☒ Cooperative activity
☒ Independent activity
☐ Concrete/manipulative activity
☒ Visual/pictorial activity
☒ Abstract procedure

Why Do It:

Each student will practice multiplication facts using a visual procedure and will gain a better conceptual understanding of these facts.

You Will Need:

A supply of multiplication charts (reproducibles are provided) and highlighter pens or colored pencils are required.

How To Do It:

This activity will allow a student to find the answer to a multiplication problem by shading a fact chart. Have students use a multiplication chart and highlighter pens or colored pencils to shade areas that show the answers to multiplication facts. On a separate piece of paper, students then write the problems and answers that the highlighted areas show.

Example:

This chart shows 3×5 (as 3 groups of 5) $= 15$. Or, viewed from the side, the chart depicts 5×3 (as 5 groups of 3) $= 15$.

10	10	20	30	40	50	60	70	80	90	100
9	9	18	27	36	45	54	63	72	81	90
8	8	16	24	32	40	48	56	64	72	80
7	7	14	21	28	35	42	49	56	63	70
6	6	12	18	24	30	36	42	48	54	60
5	5	10	15	20	25	30	35	40	45	50
4	4	8	12	16	20	24	28	32	36	40
3	3	6	9	12	15	18	21	24	27	30
2	2	4	6	8	10	12	14	16	18	20
1	1	2	3	4	5	6	7	8	9	10
×	1	2	3	4	5	6	7	8	9	10

Extensions:

Have students complete the following sample problems.

1. Shade in the area for 2×3. How many spaces did you highlight? Thus $2 \times 3 =$ ____. Explain how you have also shown the area for 3×2.

2. Highlight 3×5. Thus $3 \times 5 =$ ____ and $5 \times 3 =$ ____.

3. Show 5×8. Thus $5 \times 8 =$ ____ and $8 \times 5 =$ ____.

4. When you highlight 9×3 or 3×9, the area equals ____.

5. 7×9 or $9 \times 7 =$ ____. What do you notice, on the multiplication chart, about the location of the answer number?

6. Complete a series of highlighted charts and post them next to one another on a bulletin board. For example, do the 7s facts by highlighting 7×1 on the first chart, 7×2 on the second, 7×3 on the third, and so on.

Multiplication Charts

10										
9										
8										
7										
6										
5										
4										
3										
2										
1										
×	1	2	3	4	5	6	7	8	9	10

10	10	20	30	40	50	60	70	80	90	100
9	9	18	27	36	45	54	63	72	81	90
8	8	16	24	32	40	48	56	64	72	80
7	7	14	21	28	35	42	49	56	63	70
6	6	12	18	24	30	36	42	48	54	60
5	5	10	15	20	25	30	35	40	45	50
4	4	8	12	16	20	24	28	32	36	40
3	3	6	9	12	15	18	21	24	27	30
2	2	4	6	8	10	12	14	16	18	20
1	1	2	3	4	5	6	7	8	9	10
×	1	2	3	4	5	6	7	8	9	10

10										
9										
8										
7										
6										
5										
4										
3										
2										
1										
×	1	2	3	4	5	6	7	8	9	10

10	10	20	30	40	50	60	70	80	90	100
9	9	18	27	36	45	54	63	72	81	90
8	8	16	24	32	40	48	56	64	72	80
7	7	14	21	28	35	42	49	56	63	70
6	6	12	18	24	30	36	42	48	54	60
5	5	10	15	20	25	30	35	40	45	50
4	4	8	12	16	20	24	28	32	36	40
3	3	6	9	12	15	18	21	24	27	30
2	2	4	6	8	10	12	14	16	18	20
1	1	2	3	4	5	6	7	8	9	10
×	1	2	3	4	5	6	7	8	9	10

Computation Connections

Chapter 38

Chalkboard or Tabletop Spinner Games

Grades 2–8

☒ Total group activity

☒ Cooperative activity

☐ Independent activity

☐ Concrete/manipulative activity

☐ Visual/pictorial activity

☒ Abstract procedure

Why Do It:

Students will be able to practice with most number and operation concepts, all within an activity using a spinner to generate problems.

You Will Need:

This activity requires a large spinner, made from a piece of heavy poster board (approximately 3 inches by 18 inches); a suction cup; a short bolt; a large paper clip; and pencil and paper.

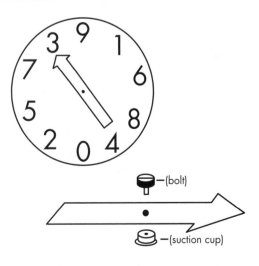

How To Do It:

1. First draw a large circle on the chalkboard with the center clearly marked with a dot. An easy way to do this is to use a string. Place one end of the string at a point (the center) on the chalkboard and hold it there, then wrap the other end of the string around the chalk and with the string taut draw a circle with the chalk. Cut the poster board into an arrow shape, punch a small hole in the arrow at the balance point, and insert the bolt through it into the suction cup. Suction the arrow to the chalkboard at the center of the circle and spin it several times to check the balance; if it needs adjustment, attach the paper clip to the arrow and move it until spins are even. Number the edge of the circle with numbers 0 through 9, as shown on page 139.

   An alternate way to do this activity is to give each individual or group of players pencils, paper, and a single paper clip. Then provide or have students draw a spinner chart with numbers 0 through 9. A student places the paper clip flat such that one end overlaps with the center of the chart, and uses a pencil point to hold the paper clip in place, flipping the paper clip with his or her other hand. The paper clip spinner will randomly point to different numbers. A sample paper clip spinner is provided after the Extensions.

2. To begin play, indicate the type of problem to be solved and have each student sketch the needed numeral blanks on his or her individual paper. For example, a place value problem in the 10,000s would require that students begin with ____ ____ , ____ ____ ____ . Say, "Try to make a number nearest to 10,000." For each spin of the spinner, students must write a number in one of the blanks on their recording sheets. Once a number has been written, it may not be moved. When all blanks are filled, the student or students with the closest answer wins.

Examples:

1. During this place value game, the teacher has stated that the answer nearest to 3,000 would win the round. Four spins yielded the numbers 5, 8, 2, and 4. Three players got the answers shown below. The winner of this round was Player 2 because he or she was closest to 3,000.

5, 2 4 8   2, 4 5 8   4, 2 5 8

(Player 1)   (Player 2)   (Player 3)

2. The object of this multiplication game was to achieve the least product. Five spins gave the numbers 9, 4, 3, 1, and 7 in that order. Three players' problems are shown below. Because the answers 14,993, 34,049, and 13,871 come up, Player 3 is the winner.

$$\begin{array}{r} 3\ 1\ 9 \\ \times\ 4\ 7 \end{array} \qquad \begin{array}{r} 4\ 3\ 1 \\ \times\ 7\ 9 \end{array} \qquad \begin{array}{r} 1\ 4\ 3 \\ \times\ 9\ 7 \end{array}$$

(Player 1)   (Player 2)   (Player 3)

Extensions:

Challenge students with a variety of question types.

1. Young players might simply be asked to name the numeral that the spinner stopped at, bounce a ball that many times, or display that number of objects.

2. Winning players might have the greatest or least results, or those nearest a designated number.

3. Organize the numeral blanks for use with fraction or decimal problems.

4. Have students try some mixed practice problems, such as ____ × ____ ÷ ____ + ____ − ____ = ____. (*Hint:* Be sure learners use the correct order of operations.)

5. Make use of some nonstandard situations. For instance, the winner might be the player who obtains the greatest remainder when dividing a 3-digit number by a 2-digit number. This is nonstandard because students are usually asked to find the quotient of a division problem as opposed to the remainder. To extend this idea further students could be asked what type of division problem would require looking at the remainder. One such example might be, "An insurance company says that you have 78 hours from January 1st at 5 P.M. to call before a claim is rendered inactive. What time on January 3rd is the last time you can call?" Students would need to divide 78 by 24 and then look at the remainder of 6 to find the time. They would find that 6 hours past 5 P.M. is 11 P.M., and therefore the call should be made by 11 P.M. on January 3rd.

6. This activity can be used to "Spin Your Own Homework." Spin five numbers and have the students write the numbers down on a piece of paper. Then instruct them to take the numbers home and put them into the following problems to get the greatest possible sum, greatest possible difference, least possible product, and quotient nearest to 50 as possible.

Problems:

$$\begin{array}{r} \underline{}\ \underline{}\ \underline{} \\ + \ \underline{}\ \underline{} \\ \hline \end{array} \qquad \begin{array}{r} \underline{}\ \underline{}\ \underline{} \\ - \ \underline{}\ \underline{} \\ \hline \end{array}$$

$$\begin{array}{r} \underline{}\ \underline{}\ \underline{} \\ \times \ \underline{}\ \underline{} \\ \hline \end{array} \qquad \underline{}\ \underline{}\ )\overline{\ \underline{}\ \underline{}\ \underline{}\ }$$

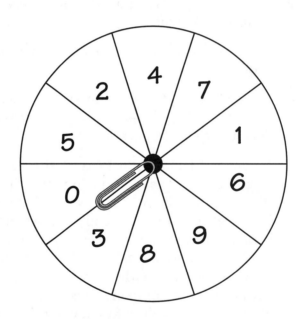

Computation Connections

Chapter 39

Skunk

Grades 3–8

☒ Total group activity
☒ Cooperative activity
☐ Independent activity
☐ Concrete/manipulative activity
☐ Visual/pictorial activity
☒ Abstract procedure

Why Do It:

This is a highly motivational game that involves both chance and computation practice.

You Will Need:

A pair of dice and one record-keeping chart per student are needed (reproducibles are provided or can be drawn as shown below).

How To Do It:

1. This is an activity where students play against each other by adding numbers and keeping score on the Skunk chart. Two dice are tossed, and if a 1 shows up it's a SKUNK. Students have fun taking chances and trying not to get SKUNKED.

   To begin, roll both dice and have all players mentally add the values together for a total that they record above the dotted line (first row below the word SKUNK) in the S columns of their SKUNK charts. (Scores above the dotted line in any column are "temporarily safe."). Then ask who would like to try for additional points in the S column; those who do are to raise their hands high, whereas any players wishing to stay out for the rest of the column lay pencils down and fold their hands. Those trying for additional points are taking a chance of getting SKUNKED, which means losing all of their points in that column if one of the dice rolled is a 1. (*Note:* A 1 rolled on the first roll of the dice does not make a SKUNK.) Play continues in the S column until no player is brave enough to try for additional points (players may drop out after any "safe" roll, but then cannot return to the game until play in the next column is started), or until a SKUNK occurs. At this time each participant totals his or her score and records it at the bottom of the S column; of course, those who got SKUNKED get zero points.

2. Play continues in the same manner for the remaining columns. A final rule, which affects scoring beginning with the second column, is the Double SKUNK rule, which states that if you are trying for additional points (below a dotted line) in any column and 1s turn up on both dice, you lose all of your scores up to that point. After the game has progressed through all of the columns, the column scores are added together for a grand total; the player with the greatest grand total is the winner.

Example:

Two players' scores in the first column of the game are shown below.

Player 1

S	K	U	N	K	
8					
6					
(Stop)					
14					

Numbers Rolled and Scores

$3 + 5 = 8$

$4 + 2 = 6$

$6 + 5 = 11$

$5 + 4 = 9$

$1 + 3 = 4$

Player 2

S	K	U	N	K	
8					
6					
11					
9					
SKUNK					
0					

Extensions:

1. Paste blank stickers on the dice and write two- or three-digit numbers on them; let numbers with repeating digits (such as 77 or 333) be the SKUNK numbers. Play the game using addition.

2. Have students play "Subtraction Skunk," in which they might, for example, subtract the smaller rolled number from the larger and record the difference on their score charts. The winner is the player nearest to an agreed-on number.

3. It might also be possible to play "Multiplication Skunk" or "Division Skunk." (In "Division Skunk," for example, students might add only the remainders to see who obtains the greatest or least grand total.)

4. Paste fraction or decimal stickers on the dice, decide on new rules, and have students play "Fraction Skunk."

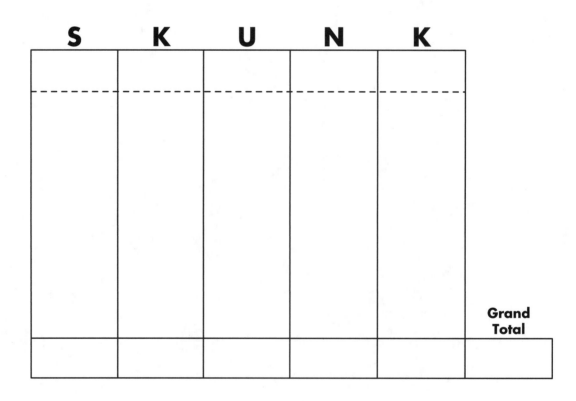

S	K	U	N	K	Grand Total

S	K	U	N	K	Grand Total

Chapter 40

Subtraction Squares

Grades 2–8

☒ Total group activity

☒ Cooperative activity

☒ Independent activity

☐ Concrete/manipulative activity

☐ Visual/pictorial activity

☒ Abstract procedure

Why Do It:

This activity allows students to practice with different forms of subtraction computation (for example, subtractions with single or multiple digits, including positive and negative numbers and zero) in a geometrical diagram.

You Will Need:

Only pencils and paper are needed. However, for some students it might be advisable to have copies of blank Subtraction Squares available; a reproducible is provided.

How To Do It:

1. In this activity, students will choose four numbers to put in the four outside corners of the diagram provided (see Subtraction Square handout). Using subtraction, they will work their way into the middle of the diagram. Students will discover a surprising end result the more Subtraction Squares they do. Using pencils and paper, the students complete the same Subtraction Square that the leader demonstrates (see Example 1) on the chalkboard or overhead projector. In Example 1, the players were asked to choose the four corner numbers to start; they selected 23, 16, 8, and 12, and placed

them in the same corners of their respective squares. They then subtracted the smaller number along each side from the larger and wrote the difference in the answer circle between them (Step 1). Next, they connected those answers with diagonal lines, subtracted again, and inserted the new answers in the middle circle on that diagonal (Step 2). They connected these new answers, vertically and horizontally, and subtracted again (Step 3). The players continued this process until no further subtracting could be done (Step 4).

2. After completing one or two such Subtraction Squares, the players will likely ask whether the result will always be zero or whether the number of squares or steps will always be the same. Suggest that they try several problems and then share what they find. (*Note:* It is advisable to have two or more students work on the same Subtraction Square, because, by doing so, they can easily compare answers and locate any discrepancies.

Examples:

1. Students selected corner numbers 23, 16, 8, and 12 for this Subtraction Square.

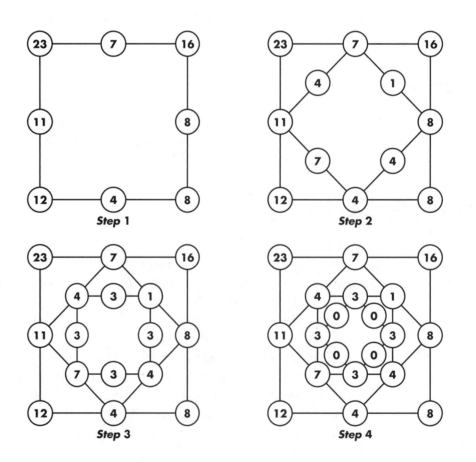

Step 1   Step 2   Step 3   Step 4

2. The corner numbers were 16, 2, −11, and 0 for this Subtraction Square.

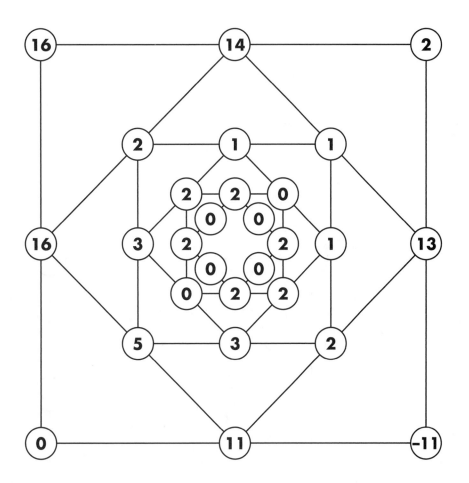

Extensions:

1. When appropriate, use Subtraction Squares as homework, instead of textbook assignments or worksheets.

2. Young players will likely be most successful when using printed Subtraction Squares (see reproducible handout) and single-digit numbers.

3. Players might attempt a variety of arrangements. For instance, try a single-digit number in one corner, a two-digit number in another, a three-digit number in the third corner, and a four-digit choice in the last.

4. Able players might be able to complete Subtraction Squares that incorporate fractions or decimals.

5. Advanced players might try to complete Division Squares. Notice that the final answers will not be zero, and that it will sometimes be necessary to round off the decimal portions of answers.

Subtraction Square

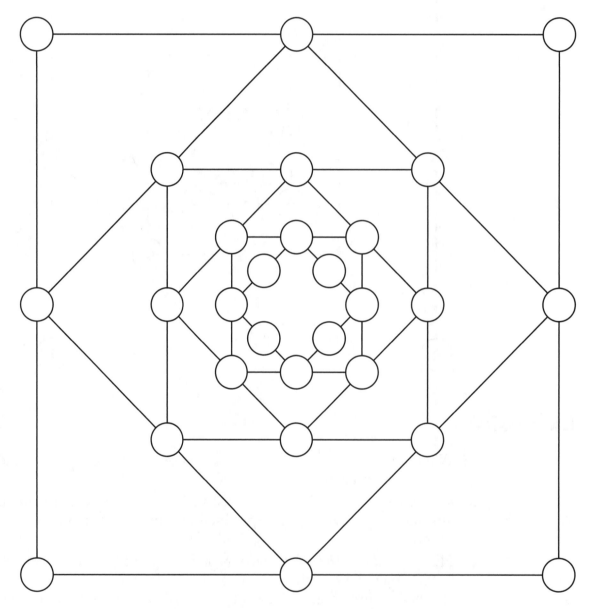

Chapter 41

Drawing Fraction Common Denominators

Grades 3–8

☒ Total group activity
☒ Cooperative activity
☒ Independent activity
☐ Concrete/manipulative activity
☒ Visual/pictorial activity
☒ Abstract procedure

Why Do It:

This activity helps students gain a visual perspective of what a common denominator is and why it is used.

You Will Need:

Graph paper or the reproducibles provided in *Number Cutouts* (p. 22) are required. Students can also simply draw the rectangles on a plain sheet of paper.

How To Do It:

In this activity, students will use graph paper or draw their own grids to show fraction common denominators. The "secret" is to use the denominator (bottom) numbers in any pair of fractions and to draw a rectangle that has a length equal to one of the denominators and a width equal to the other denominator. For example, if the fractions are 1/3 and 1/4, as in Example 1, students must draw a rectangle that is 3 units by 4 units, which shows 12 square units.

Examples:

Guide students through the following examples before having them attempt the Extensions.

1. For $1/3 + 1/4 =$ _____, have the students draw three rectangles that are 3 units by 4 units, or 12 square units. Have them shade 1/3 of the total area of one rectangle, which is 4/12; and shade 1/4 of another rectangle, which is 3/12. Next, they draw the 4/12 and the 3/12 onto the third rectangle grid for a total of 7/12. (*Note:* When students draw the 4/12 and 3/12 onto one grid, they should not overlap the shading.)

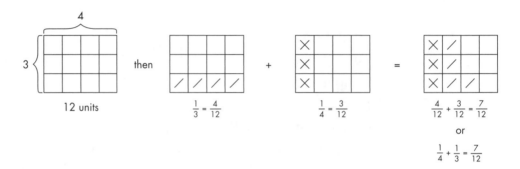

2. For a slightly harder problem, guide students through the steps for the problem $1/6 + 2/3 =$ _____. First draw four rectangles that are 6 units by 3 units, or 18 square units. Shade 1/6 of one rectangle, which is 3/18; and shade 2/3 of another rectangle, which is 12/18. Combine these to get 15/18, which is correct, but not in lowest terms.

   To reduce the 15/18 to lowest terms, look for ways to group units; in this case, both the 15 and the 18 can be grouped into 3s. Draw loops to show the 15 as 5 groups of 3 and the 18 (the entire rectangular grid) as 6 groups of 3. Out of the 6 loops, 5 are shaded. Therefore, 15/18 can be renamed and reduced to 5/6.

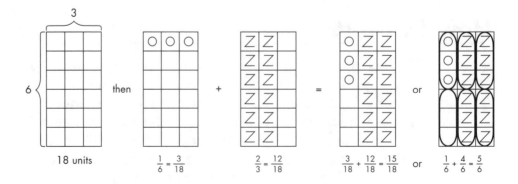

Computation Connections

Extensions:

You can extend this activity with the following problems.

1. Draw diagrams for $1/2 + 1/3 =$ ___.
2. Show $2/5 + 1/2 =$ ___.
3. Show $1/2 + 1/6 =$ ___ in lowest terms. (*Hint:* Use groups of 4.)
4. Show $1/4 + 1/6 =$ ___ in lowest terms.
5. Try several problems of your own or "prove" the results to some assigned fraction problems.

Chapter 42

Fraction × and ÷ Diagrams

Grades 4–8

☒ Total group activity

☒ Cooperative activity

☒ Independent activity

☐ Concrete/manipulative activity

☒ Visual/pictorial activity

☒ Abstract procedure

Why Do It:

This activity helps students gain a visual perspective of what happens when fractions are multiplied and divided.

You Will Need:

Obtain a supply of graph paper (or make duplicate copies from the *Number Cutouts* activity, p. 22) and distribute two or three sheets to each student. Colored pencils are optional, but will help to make the student diagrams more appealing. The first time this activity is presented to the class, consider using large, demonstration-size graph paper, or an overhead projector and colored pens.

How To Do It:

1. In this activity, students will use graph paper and the shading of rectangles to picture multiplication and division of fractions. Often students just memorize the rules for these operations, but never know why these rules work.

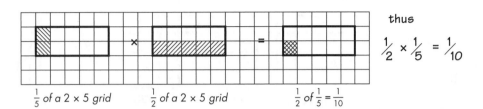

thus

$$\frac{1}{2} \times \frac{1}{5} = \frac{1}{10}$$

$\frac{1}{5}$ of a 2 × 5 grid $\frac{1}{2}$ of a 2 × 5 grid $\frac{1}{2}$ of $\frac{1}{5} = \frac{1}{10}$

Use the large graph paper or an overhead projector to work through several problems with the class. For example, with the problem $1/2 \times 1/5 = $ ____ (illustrated above), begin by helping students understand that because the denominators are 2 and 5, they will need to utilize a 2-unit by 5-unit grid. Then shade the first column and mark it 1/5, as it is 1/5 of the whole rectangle. Next shade the bottom row and mark it 1/2, because it is 1/2 of the whole rectangle. Finally, note where these shaded areas overlap and find the fraction that represents this overlapped region. In this problem, the fractional answer is 1/10, and therefore 1/2 of the 1/5 portion equals 1/10 of the entire rectangle. It has thus been shown that $1/2 \times 1/5 = 1/10$. (*Note:* Some students find it helpful to think of this situation as 1/2 of 1/5 = 1/10.)

2. A similar process is followed when dividing a fraction by a fraction, as demonstrated in the Example below. Finally, the slightly more complex issue of multiplying (or dividing) a mixed number by a mixed number is illustrated in the Extensions section. After demonstrating as many examples as necessary, have students try some problems on their own.

Example:

The students depicted here are working cooperatively on drawing a diagram for $1/4 \div 1/3 = $ ____. The area of the first shaded region is 3 square units, and the area of the second shaded region is 4 square units, so the answer is 3/4. Notice the students' thoughts about the division process, including the application the last student came up with.

Extensions:

When they are ready, students might be asked to visualize and diagram mixed number situations, such as those that follow.

1. The diagram below shows a means of representing the problem 1-1/3 × 2-1/4 = _____.

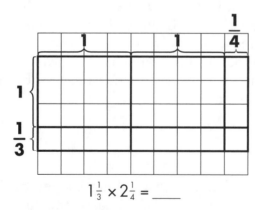

$1\frac{1}{3} \times 2\frac{1}{4} =$ _____

$\frac{4}{3} \times \frac{9}{4} = \frac{36}{12}$ or 3 (whole areas)

(Note: 1 whole area is 3 × 4 or 12 spaces)

2. Applying the numerals used in Extension 1 to the process of division yields quite a different result. In this situation, 1-1/3 ÷ 2-1/4 = _____ might be illustrated as shown below.

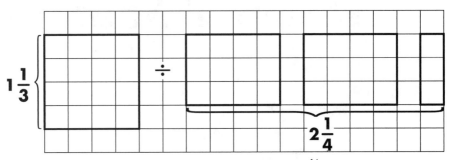

(16 squares) ÷ (27 squares) = $\frac{16}{27}$

Chapter 43

Decimal Squares

Grades 2–8

☒ Total group activity
☒ Cooperative activity
☒ Independent activity
☐ Concrete/manipulative activity
☒ Visual/pictorial activity
☒ Abstract procedure

Why Do It:

Students will compare decimals and compute problems with decimals using a visual model.

You Will Need:

A supply of Decimal Squares (reproducible is provided), highlighter pens, and colored pencils or crayons are required.

How To Do It:

1. This activity will make use of Decimals Squares to compare, add, subtract, and multiply decimal numbers. Each Decimal Square is a 10-unit by 10-unit square; in other words, it has an area of 100 square units. Each small square represents one hundredth of the Decimal Square, or .01. Each column (of 10 squares) represents one-tenth, or 0.1, of the Decimal Square.

   Start by shading 0.2 (two columns of 10 square units) on a Decimal Square (using an overhead projection system, if possible) and explain that it represents two-tenths of the whole square. Then shade 0.32 (three columns and 2 square units) on a Decimal Square and explain that it represents thirty-two hundredths.

2. Next use a Decimal Square or squares and colored writing utensils to shade areas that show the problem, and discuss how to find the answer to the problem using the shaded squares. Below each Decimal Square, show the problem and the solution to the problem. The solution should be an inequality sign or a decimal number. See the Examples below to use as demonstrations. (*Note*: When designing problems for this activity, be careful that the answers do not include the thousandth place, or the problems will not be possible to solve using a Decimal Square.)

Examples:

The following problems can be used to explain the use of Decimal Squares to students.

1. Shade 0.6 in the first Decimal Square (starting on the left) and 0.21 in the second (starting on the left). Then compare the shaded areas to discover the answer to the problem, "To solve 0.6____ 0.21, fill in the blank with >, <, = ."

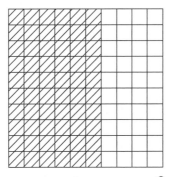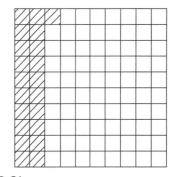

0.6 > 0.21

2. Shade 0.25 using one color and 0.3 using another color. Add up the entire area to discover the sum of 0.25 and 0.3.

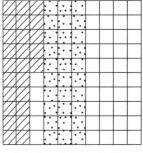

0.25 + 0.3 = 0.55

Decimal Squares

3. Shade 0.46 in one color, then shade 0.2 in the area already shaded using another color. The part of the figure shaded twice will be subtracted from the total area. Ask students what decimal represents the area remaining.

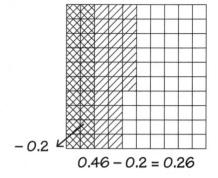

-0.2

$0.46 - 0.2 = 0.26$

4. Shade 0.5 vertically and 0.4 horizontally. The number of squares in the rectangular area of overlap represent the solution to the problem 0.5×0.4.

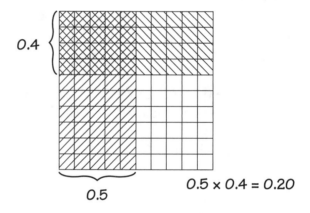

0.4

0.5

$0.5 \times 0.4 = 0.20$

Extensions:

Have students try the following problems using Decimal Squares.

1. Insert $<$, $>$, or $=$ to make the statement 0.34 ____ 0.3 true.

2. Insert $<$, $>$, or $=$ to make the statement 0.5 ____ 0.50 true.

3. Find $0.4 + 0.27$.

4. Find $0.72 + 0.15$.

5. Find $0.54 - 0.3$.

6. Find $0.68 - 0.4$.

7. Find 0.5×0.6.

8. Find $0.3 - 0.8$.

9. Using the Decimal Square below, state the decimal multiplication problem represented by the rectangle shaded.

Rectangle A → *0.4 × 0.5 = .20*

Rectangle B →

Rectangle C →

Rectangle D →

Rectangle E →

Rectangle F →

Rectangle G →

Rectangle H →

Rectangle I →

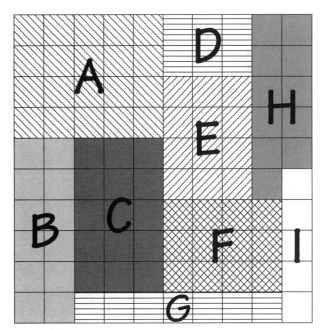

Decimal Squares

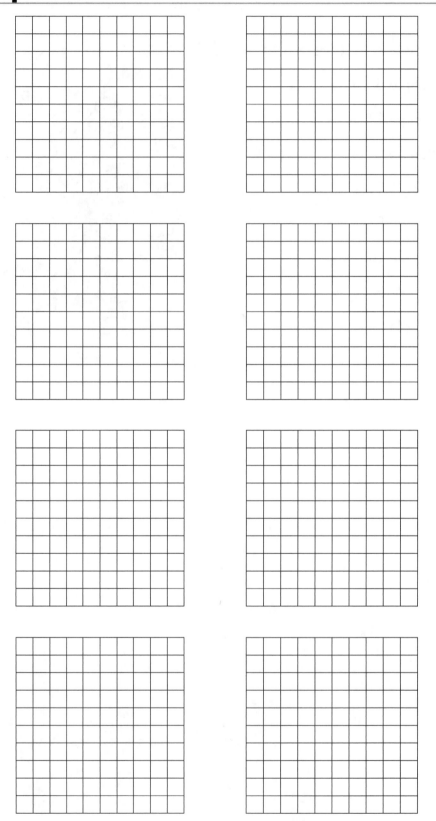

Computation Connections

Chapter 44

Square Scores

Grades 2–8

☐ Total group activity
☒ Cooperative activity
☐ Independent activity
☐ Concrete/manipulative activity
☒ Visual/pictorial activity
☒ Abstract procedure

Why Do It:

Students will practice with addition, subtraction, multiplication, or division facts, using logical-thinking strategies in a game setting.

You Will Need:

Square Score Grids (provided at the end of this activity) are required. Usually one per pair of students is enough to start with. Once they are familiar with the activity, players might also devise grids for each other (see Extensions). Pencils and pens of different colors are also needed.

How To Do It:

Square Scores is usually played by two students on one grid. The grid contains 5 rows and 7 columns of dots. In the middle of a group of four adjacent dots is a math problem. Each student uses a pencil or pen of a different color, and at her or his turn draws a vertical or horizontal line between any two adjacent dots. Play continues in this manner until a line is drawn that closes a square. The student who draws that line must attempt to answer the problem contained within that box. If the problem is answered correctly, that student is allowed to claim the square and to shade or mark it. If the student gives an incorrect answer, the square is marked with an X and no credit is allowed. (Students might check their answers with a calculator or an answer sheet.) When all squares are closed, the students count the boxes claimed to see how many facts they knew.

Example:

The players pictured below are practicing their multiplication facts for 6s, while also attempting to capture as many squares as possible. Thus far Juanita has captured and marked the three squares marked \\\\, and Jose has claimed the two facts marked ////.

Extensions:

1. If students need practice with a certain operation, such as subtraction, then the grid should utilize only those types of problems. However, if mixed practice is desirable, a different grid might include a combination of addition, subtraction, multiplication, or division or even fractions or decimals.

2. *Square Scores* also works well as a team game when it is played on the overhead projector. In such a setting, the team members are allowed a strategy conference (for two minutes), and then the team leader draws the line for that turn. Play continues in this manner until all squares on the overhead transparency are surrounded and marked. The winning team is the one that has captured the most squares.

3. Players can easily devise their own grids by writing equations designated for practice on blank grids (see model provided) or by using one-inch or larger graph paper. (*Note:* The grid designer should also create an answer key.) The designed grid can be photocopied and tried by several other players.

4. Advanced levels of the game might include having three, four, or more players competing on the same grid, and could include bonus squares (enclosing problems more difficult than those typical for the grade or age level.

SQUARE SCORE ADDITION AND SUBTRACTION

2 + 3	5 + 1	6 − 2	4 − 4	3 + 7	1 + 4
5 + 5	6 − 4	2 + 2	8 + 2	6 − 3	3 + 3
7 − 4	5 + 6	4 + 4	7 − 5	3 + 6	4 + 2
2 + 5	9 − 5	6 + 6	5 + 5	3 + 8	9 − 2

SQUARE SCORE MULTIPLICATION AND DIVISION

5×5 $15 \div 3$ 4×6 $12 \div 3$ 7×4 6×2

2×7 5×8 $27 \div 9$ 6×6 $18 \div 2$ 9×7

$45 \div 9$ 7×8 $35 \div 7$ $36 \div 4$ 6×9 $24 \div 8$

$9 \div 3$ 9×7 6×8 $60 \div 10$ 3×8 9×2

CREATE YOUR OWN SQUARE SCORE GRID

Square Scores

Chapter 45

Math Concentration

Grades 2–8

☒ Total group activity
☒ Cooperative activity
☐ Independent activity
☐ Concrete/manipulative activity
☒ Visual/pictorial activity
☒ Abstract procedure

Why Do It:

Students will make connections between mathematical terms, numbers, basic facts, and geometric figures.

You Will Need:

Small cards (3- by 5-inch index cards, for example) and marking pens are required.

How To Do It:

Give each player two or four cards. Each player will then select a math fact or concept, such as 3×4, and on their cards will represent this fact in different ways. When all the players' cards are ready, they are shuffled together and placed face down in rows and columns. A player then turns up two cards; if the player can "prove" a match, he or she gets to keep the cards and try again. If not, the cards are returned to their spots, and *Math Concentration* continues with the next player. The player holding the most cards at the end of the game wins.

Examples:

1. These cards match the two concepts *perpendicular* and *intersecting* with a drawing.

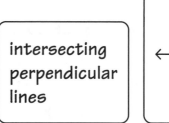 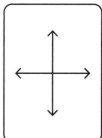

2. A dot diagram, an addition problem, a multiplication fact, and the numeral itself have been used as four ways to represent 20 for the game depicted below.

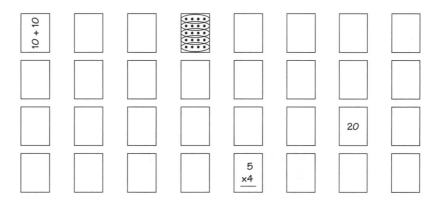

Extensions:

1. Students can match equivalent fractions, fractions with decimals, fractions or decimals with percentages, and so on.

2. Together with either English or metric rulers, students can match measurements with pictures or drawings of corresponding lengths (for example, pairing 2 inches with a circle of that diameter).

3. Simple word problems and solutions can be paired. "Tricky" problems, such as the following, can be fun too: "How much dirt can be taken from a hole 6 feet long by 2 feet wide by 1 foot deep? *Answer:* NONE! *Reason:* It is a hole, so the dirt is already gone.

Chapter 46

Scramble

Grades 2–8

☒ Total group activity
☒ Cooperative activity
☐ Independent activity
☐ Concrete/manipulative activity
☐ Visual/pictorial activity
☒ Abstract procedure

Why Do It:

This activity helps reinforce students' grasp of addition, subtraction, multiplication, and division facts, and requires them to practice mental and pencil-and-paper computation.

You Will Need:

Copies of Scramble Cards (reproducibles provided) are required.

How To Do It:

Scramble is usually played with two to four teams of ten people. Each team member holds a single, colored card with a numeral from 0 to 9 on it; the cards for one team might be red, another team green, and so on. The teacher calls out a number problem and the students from each team who are holding the correct answer numerals "scramble" (walk or run) to the answer area for their team. The answer area might be in front of the classroom or in opposite corners of the classroom, or even in designated places outside. The teacher at some point calls out "Freeze," and at this point students have to freeze in their positions. Each team achieving a correct answer receives a point, and the first team to do so is also given a bonus point. In the case of a tie, each tying team receives one bonus point. The team with the highest score wins.

Example:

In the situation shown below, the leader has called out 8 + 9. The Red Team players with the 1 and the 7 have scrambled to their answer location to show 17 as the proper answer, and have received a point for the correct answer plus a bonus point for being first. The Green Team has the correct numerals, but in the wrong order; if they reorder before the leader says "Freeze," they might still get one point.

RED TEAM = WINNER   GREEN TEAM = WRONG ORDER

Extensions:

If simple questions are called out, it is quickest for students to do these in their heads. For more difficult ones, the teams may talk them through, use pencil and paper to help find answers, or use calculators. Have students try some of the following problems (or any other appropriate problems), and allow them to design problems of their own.

1. Addition:

Basic facts	3 + 4 = ____	5 + 5 = ____	7 + 8 = ____
No carrying	12 + 12 = ____	33 + 11 + 21 = ____	132 + 246 = ____
With carrying	8 + 13 = ____	25 + 26 = ____	98 + 103 = ____

2. Subtraction:

Basic facts	7 − 2 = ____	12 − 9 = ____	13 − 5 = ____
No borrowing	24 − 12 = ____	46 − 31 = ____	147 − 22 = ____
With borrowing	33 − 24 = ____	87 − 19 = ____	312 − 215 = ____

3. Other:

Multiplication or division	7 × 9 = ____	96 ÷ 12 = ____
Mixed	(5 × 8) + (the days in a week) − 13 = ____	
Nontraditional	(Only show the remainder from 108 ÷ 12)	

SCRAMBLE CARDS

4	3	2	1	0
9	8	7	6	5

Chapter 47

Palindromic Addition

Grades 2–8

☒ Total group activity

☒ Cooperative activity

☒ Independent activity

☐ Concrete/manipulative activity

☐ Visual/pictorial activity

☒ Abstract procedure

Why Do It:

This activity provides a challenging and interesting way to practice addition.

You Will Need:

Although pencils and paper are the only materials required, the "Guide to Palindromic Sums" for numbers less than 1,000 (at the end of this activity) will likely prove helpful.

How To Do It:

1. To begin, demonstrate the activity to students by selecting a number less than 1,000 that is not palindromic (reversible) and adding the number to it that is obtained by reversing its digits (for example, if 158 is the selected number, add 851 to it). Continue manipulating the resulting sums in this manner until a palindromic sum (a sum that reads the same in either direction) is attained.

2. As students begin to work in small groups or individually, it is suggested that they work with 3- or 4-step solution numbers for their initial attempts (see the "Guide to Palindromic Sums"). Also, at least two students should work together on each problem so that they can compare and check their work. Players who are ready can then go on to try problems of 6, 8, 10, or even 24 steps!

Examples:

1. A palindromic sum is achieved here in 3 steps.

```
    158
   +851
   ────
   1009
  +9001
  ─────
  10010
 +01001
 ──────
  11011
```

2. Here, 6 steps are necessary.

```
     79
    +97
    ───
    176
   +671
   ────
    847
   +748
   ────
   1595
  +5951
  ─────
   7546
  +6457
  ─────
  14003
 +30041
 ──────
  44044
```

Extensions:

1. Young players who are able to add can work with numbers lower than 20.

2. Also look for words that are palindromes (for example, *mom*, *dad*, and *level*). This is not mathematical, but a way to connect what the class is doing in math to words and their spellings.

3. Find out what happens when a palindrome is added to itself in the first step (for example, $88 + 88 = 176$, and $176 + 671 = 847$, and $847 + 748 =$ ____, and so on).

4. Challenge students to find a palindromic sum in exactly 7 steps, for example, or to discover a palindromic sum having more than 13 digits (see 89 or 98 in the "Guide to Palindromic Sums").

5. Have students determine color-coded patterns for the numbers from 1 to 100 on either 99s or 100s charts, as printed below. Enlarge and duplicate copies of the following charts, allow the players to select their own color schemes, and have them individually or in small groups shade in the patterns on the charts.

Choose a color for each

0	1	2	3	4	5	6	7	8	9
10	11	12	13	14	15	16	17	18	19
20	21	22	23	24	25	26	27	28	29
30	31	32	33	34	35	36	37	38	39
40	41	42	43	44	45	46	47	48	49
50	51	52	53	54	55	56	57	58	59
60	61	62	63	64	65	66	67	68	69
70	71	72	73	74	75	76	77	78	79
80	81	82	83	84	85	86	87	88	89
90	91	92	93	94	95	96	97	98	99

Try a 99s Chart

Choose a color for each

1	2	3	4	5	6	7	8	9	10
11	12	13	14	15	16	17	18	19	20
21	22	23	24	25	26	27	28	29	30
31	32	33	34	35	36	37	38	39	40
41	42	43	44	45	46	47	48	49	50
51	52	53	54	55	56	57	58	59	60
61	62	63	64	65	66	67	68	69	70
71	72	73	74	75	76	77	78	79	80
81	82	83	84	85	86	87	88	89	90
91	92	93	94	95	96	97	98	99	100

Try a 100s Chart

Guide to Palindromic Sums

	"sum"	numbers
3 steps	11,011	158, 257, 356, 455, 544, 653, 752, 851, 950
	13,431	168, 267, 366, 465, 564, 663, 762, 861, 960
	15,851	178, 277, 376, 475, 574, 673, 772, 871, 970
	3,113	199, 298, 397, 496, 694, 793, 892, 991
	5,115	249, 348, 447, 546, 645, 744, 843, 942
	5,335	299, 398, 497, 596, 695, 794, 893, 992
	6,666	156, 255, 354, 453, 552, 651, 750
	8,888	157, 256, 355, 553, 652, 751, 850
	6,996	186, 285, 384, 483, 582, 681, 780
	7,337	349, 448, 547, 745, 844, 943
	7,117	389, 488, 587, 785, 884, 983
	7,557	399, 498, 597, 795, 894, 993
	9,119	439, 538, 637, 736, 835, 934
	9,559	449, 548, 647, 746, 845, 944
	9,339	489, 588, 687, 786, 885, 984
	9,779	499, 598, 697, 796, 895, 994
	4,444	155, 254, 452, 551, 650
	2,662	164, 263, 362, 461, 560
	4,884	165, 264, 462, 561, 660
	2,552	184, 283, 382, 481, 580
	4,774	185, 284, 482, 581, 680
	2,992	194, 293, 392, 491, 590
	1,111	59, 68, 86, 95
	747	180
4 steps	5,115	174, 273, 372, 471, 570
	9,559	175, 274, 472, 571, 670
	9,339	195, 294, 492, 591, 690
	4,884	69, 78, 87, 96
	25,652	539, 638, 836, 935
	23,232	579, 678, 876, 975
	22,022	599, 698, 896, 995
	45,254	629, 728, 827, 926
	44,044	649, 748, 847, 946
	47,674	679, 778, 877, 976
	46,464	699, 798, 897, 996
	13,431	183, 381, 480
	6,996	192, 291, 390
	69,696	729, 927
	68,486	749, 947
	67,276	769, 967
	66,066	789, 987
	89,298	819, 918
	88,088	839, 938
	2,662	280
	2,5552	290

(Continued)

Guide to Palindromic Sums (continued)

	"sum"	*numbers*
5 steps	79,497	198, 297, 396, 495, 594, 693, 792, 891, 990
	45,254	166, 265, 364, 463, 562, 661, 760
	44,044	176, 275, 374, 473, 572, 671, 770
	59,895	549, 648, 846, 945
	99,099	639, 738, 837, 936
6 steps	45,254	182, 281, 380
	44,044	79, 97
	475,574	779, 977
	449,944	799, 997
	881,188	889, 988
7 steps	233,332	188, 287, 386, 485, 584, 683, 782, 881, 980
	881,188	197, 296, 395, 593, 692, 791,890
	45,254	190
8 steps	1,136,311	589, 688, 886, 985
	233,332	193, 391, 490
10 steps	88,555,588	829,928
11 steps	88,555,588	167, 266, 365, 563, 662, 761, 860
14 steps	8,836,886,388	849, 948
15 steps	8,836,886,388	177, 276, 375, 573, 672, 771, 870
17 steps	5,233,333,325	739, 937
	133,697,796,331	899, 998
22 steps	8,813,200,023,188	869, 968
23 steps	8,813,200,023,188	187, 286, 385, 583, 682, 781, 880
	8,802,236,322,088	879, 978
24 steps	8,813,200,023,188	89, 98

Chapter 48

Paper Clip Division

Grades 3–8

☒ Total group activity
☒ Cooperative activity
☒ Independent activity
☒ Concrete/manipulative activity
☒ Visual/pictorial activity
☒ Abstract procedure

Why Do It:

Students will have a concrete, 1-to-1 experience as they solve division problems. This experience will help them internalize the meaning of division.

You Will Need:

This activity requires one or more boxes of paper clips, and pencils and paper.

How To Do It:

Have students separate groups of paper clips in order to practice division, beginning with the Examples below. Also have them use pencil and paper to record their results.

Examples:

Guide students through the following problems.

1. For 12 ÷ 2, have students take 12 paper clips and divide them into 6 groups of 2. In this case, they should

take 2 paper clips out of 12 and set them aside, then keep doing this until they cannot take 2 anymore. The number of groups of 2 is the answer, which is 6, and because there are no leftover clips, there is no remainder.

$$\text{(shows } 2\overline{)12}\text{)}$$
$$\begin{array}{r} 6 \\ 2\overline{)12} \\ 12 \\ \hline 0 \end{array}$$

2. For $44 \div 7$, have students use 44 paper clips and put them into groups of 7. You will get 6 groups of 7, plus 2 extra paper clips.

$$\text{(shows } 7\overline{)44}\text{)}$$
$$\begin{array}{r} 6 \\ 7\overline{)44} \\ 42 \\ \hline 2 \end{array}$$

Extensions:

Students should work with paper clips and then with numbers to complete the following problems. Have them share their findings with the group, the leader, or other players.

1. Count out 25 clips and divide them into groups of 5. → $5\overline{)25}$.

2. Show $21 \div 6$ with clips. → $6\overline{)21}$

3. Show $9\overline{)63}$ with clips and then with numbers.

4. Use 2 boxes of paper clips to show 198 divided by 37. → $37\overline{)198}$

5. Examples 1–4 all involved measurement division (when you know the number in a set but not the number of sets). Also try some problems concerning partitive division (when you know the number of sets but not the number in each set). The following problems illustrate the difference.

$$\begin{array}{r} 6 \text{ people get apples} \\ 2 \text{ each } \overline{)12} \text{ apples} \\ \underline{12} \\ 0 \end{array}$$

(Measurement Division)

$$\begin{array}{r} 6 \text{ apples per person} \\ 2 \text{ people } \overline{)12} \text{ apples} \\ \underline{12} \\ 0 \end{array}$$

(Partitive Division)

6. Make up some paper clip problems of your own and share them with the class.

Chapter 49

I Have ___,
Who Has ___?

Grades 3–8

☒ Total group activity
☒ Cooperative activity
☐ Independent activity
☐ Concrete/manipulative activity
☐ Visual/pictorial activity
☒ Abstract procedure

Why Do It:

This activity will enhance students' mental-math computation abilities, provide focused reviews of math facts, and promote logical-thinking skills.

You Will Need:

Index cards or card stock to make playing cards and marking pens to write on the cards are required. (Reproducible handouts are also provided.)

How To Do It:

A sequential set of *I Have ____, Who Has ____?* playing cards, with one card per student, needs to be prepared in advance. (You are also provided with sample cards for thirty students, which include a variety of question types, along with a blank form for creating new cards.) The cards should be well mixed and then randomly distributed. A designated leader starts the activity by calling out the "Who has ____" (question) from his or her card. All the other players then look at the "I have ____" (answer) portions of their cards to see whether they might have the correct response. The player with the proper answer then calls out his or her "I have " (correct answer), and, if all agree, then reads aloud the "Who has ____" (new question) portion of his or her card. Play continues in this manner until each player has both correctly answered a question (assistance may be provided) and has asked a question of the other players. (*Note:* If there are exactly the same number of playing cards and players, the final answer will be on the designated leader's card. In other words, play will come back to the leader. If there are more cards than players, some players should hold two cards.)

Example:

In the situation shown here (for just four players), John, as the designated leader, called out "Who has 8 + 9?" Sara responded, "I have 17" and, after a pause to determine if all agreed, read aloud, "Who has the number of wheels on 4 tricycles?" In turn, Amber responded, "I have 12," and then called out, "Who has 6 × 4 − 5?" Jose read, "I have 19," and then asked, "Who has the number of ears on 8 students?" John stated, "I have 16." This completed the game, because John, as the leader, had asked the first question and now had answered the final question.

Extension:

The following *I Have ____, Who Has ____?* activity has been set up with relatively easy problems. Cut out the cards, distribute one to each player (if there are fewer than thirty players, some may need to hold more than one card), and allow the students to play and enjoy this sample game while also learning the procedure. Next, make copies of the blank game cards; fill in appropriate questions and responses so that students can enjoy playing while also enhancing mental-math and logical-thinking skills. (*Note:* If the students are able, challenge them to create their own cards.)

I Have _____, Who Has _____?   **183**

I Have _____ , Who Has _____?

I HAVE 100 WHO HAS 2 + 2?	I HAVE 3 WHO HAS 30 – 4?	I HAVE 27 WHO HAS the number of ears on 8 students?
I HAVE 30 WHO HAS 15 – 5?	I HAVE 11 WHO HAS 6 + 6?	I HAVE 21 WHO HAS 11 + 11?
I HAVE 8 WHO HAS the number of legs on 5 dogs?	I HAVE 15 WHO HAS 4 × 7?	I HAVE 29 WHO HAS 10 + 10 + 3?
I HAVE 24 WHO HAS the number of sides on a hexagon?	I HAVE 14 WHO HAS 5 + 5 + 3?	I HAVE 18 WHO HAS 20 – 3?
I HAVE 1 WHO HAS 30 – 28?	I HAVE 7 WHO HAS 5 x 5?	I HAVE 9 WHO HAS 20 – 1?
I HAVE 2 WHO HAS 3 + 4?	I HAVE 25 WHO HAS 3 + 3 + 3?	I HAVE 19 WHO HAS 10 × 10?
I HAVE 6 WHO HAS 7 + 7?	I HAVE 13 WHO HAS 20 – 2?	I HAVE 17 WHO HAS 100 – 99?
I HAVE 20 WHO HAS the number of wheels on 5 tricycles?	I HAVE 28 WHO HAS 30 – 1?	I HAVE 23 WHO HAS 8 + 8 + 8?
I HAVE 10 WHO HAS 5 + 6?	I HAVE 12 WHO HAS 30 – 9?	I HAVE 22 WHO HAS 5 + 3?
I HAVE 4 WHO HAS the number of sides on a triangle?	I HAVE 26 WHO HAS 9 + 9 + 9?	I HAVE 16 WHO HAS 40 – 10?

Computation Connections

I Have _____ , Who Has _____?

I HAVE WHO HAS ?	I HAVE WHO HAS ?	I HAVE WHO HAS ?
I HAVE WHO HAS ?	I HAVE WHO HAS ?	I HAVE WHO HAS ?
I HAVE WHO HAS ?	I HAVE WHO HAS ?	I HAVE WHO HAS ?
I HAVE WHO HAS ?	I HAVE WHO HAS ?	I HAVE WHO HAS ?
I HAVE WHO HAS ?	I HAVE WHO HAS ?	I HAVE WHO HAS ?
I HAVE WHO HAS ?	I HAVE WHO HAS ?	I HAVE WHO HAS ?
I HAVE WHO HAS ?	I HAVE WHO HAS ?	I HAVE WHO HAS ?
I HAVE WHO HAS ?	I HAVE WHO HAS ?	I HAVE WHO HAS ?
I HAVE WHO HAS ?	I HAVE WHO HAS ?	I HAVE WHO HAS ?
I HAVE WHO HAS ?	I HAVE WHO HAS ?	I HAVE WHO HAS ?

Chapter 50

Number Grids

Grades 3–8

☒ Total group activity
☒ Cooperative activity
☒ Independent activity
☐ Concrete/manipulative activity
☐ Visual/pictorial activity
☒ Abstract procedure

Why Do It:

This activity challenges students to locate equations that involve more than one operation, encouraging practice with addition, subtraction, multiplication, and division facts.

You Will Need:

At first students can use a prepared Number Grid (sample provided). Once they are familiar with the activity, the students might devise Number Grids for each other (see Extension 2). Also needed is a pencil for each group and colored pencils or pens.

How To Do It:

A Number Grid, a chart with 14 rows and 7 numbers in each row, is provided as a starting point for this activity. It is best to have students work in groups on the Number Grid provided. Students will use a pencil or pen to loop and label as many correct equations as possible on their grid. Using different colors is recommended if overlapping equations are allowed. The numerals for an equation must be in adjacent spaces. Give students time to work, and then discuss their findings. If desired, use an overhead transparency of the same Number Grid on which the students are working to draw attention to certain equations, in which case each student should contribute one or more looped equations to the display.

Example:

Several equations are looped on the sample Number Grid below. Notice that some involve a single operation, whereas others use several.

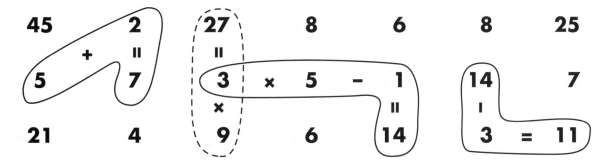

Extensions:

1. If students require practice with a certain operation, such as multiplication, ask them to loop only multiplication equations.

2. Students can easily devise a grid to be used by the class. This is easily done by placing the numerals for previously selected equations in adjacent spaces on a blank grid (graph paper works well) and then filling the remaining spaces with other numbers. Once completed, a student's grid might be photocopied and tried by several other students.

3. Advanced players can restrict themselves to finding equations that use at least three of the four basic operations.

4. A long-term game may run on for several days and involve having the players work on the grid at home. Students may be allowed to receive help from friends, parents, or anyone who wishes to contribute. In such a situation, be certain that students use different colored pens or pencils and have the equations written on additional paper for easier verification. The players may wish to check one another's work. Calculators may be helpful, because most models use the proper order of operations. Be prepared to find more than 100 equations on most one-page grids.

Number Grid

45	6	81	42	7	6	19
5	2	27	8	6	8	25
9	7	3	5	1	14	7
21	4	9	6	14	3	11
56	7	28	2	15	2	9
3	18	2	3	10	8	3
36	9	14	2	28	42	7
4	5	7	56	9	6	4
20	4	2	70	3	14	24
80	1	28	35	17	22	6
4	19	7	35	29	2	16
15	0	5	1	10	11	2
60	19	12	69	5	22	8
67	34	24	105	79	57	32

Computation Connections

Chapter 51

Here I Am

Grades 3–8

☒ Total group activity
☒ Cooperative activity
☐ Independent activity
☐ Concrete/manipulative activity
☐ Visual/pictorial activity
☒ Abstract procedure

Why Do It:

Students will reinforce their knowledge of multiplication facts, stimulate logical thinking, and enhance coordinate-graphing skills.

You Will Need:

This game requires a "master" multiplication game board with answers and several enlarged "Player Game Boards" without answers (see reproducibles). Lettered discs (circles made out of card stock, or some sort of plastic round disc that you can write on) for the message HERE I AM (or another selected short message) should be made. Furthermore, the master board (and possibly the student boards as well) may be covered with a plastic lamination or clear self-stick vinyl so that it can be written on with water-soluble marking pens or grease pencils and then erased later.

How To Do It:

1. Before play begins, put a hidden message on the round discs, one letter on each disc. The hidden message is HERE I AM in the Example below. Then tell the players what the hidden message will be and place the lettered discs on the Master Game Board, being careful to keep their locations secret from the players. For example, on the Master Game Board, the H is placed on the 21, the E is placed on the 18, and so on. The discs may be placed in horizontal, vertical, or diagonal fashion, but the letters of each word must be in adjacent spaces and there may be only one space between words. One method to keep it hidden is to put an open book on end in front of the Master Game Board.

2. The game begins when one player calls out a pair of multiplication factors and suggests an answer. If the group agrees that the stated answer is correct, each player writes it on his or her own Player Game Board in the proper position (for example, if the player notes that $7 \times 3 = 21$, the answer must be written in the location 7 over and 3 up, or $+7$ along the x-axis and $+3$ on the y-axis). Further, if the answer matches a lettered space on the Master Game Board, state whether the answer is correct or not and then say "Here I am," and all players mark that location.

3. The game continues with players, in turn, calling new pairs of factors and answers, and recording the products in the proper locations on their grids. At any turn, if a player "hits" a lettered location, say "Here I am." When one or more players think they know the location of all the HERE I AM discs (or those for another specified message), they may ask to be "checked"; at this point they must call out all of the multiplication fact problems and answers that correctly indicate the disc locations. Any player to properly do so is a winner!

Example:

The leader hid the HERE I AM discs as shown on the Master Game Board here. When the first player called out the fact that $7 \times 7 = 49$, the leader said, "Correct" (but not "Here I am," because there is no letter from the clue at that location), so all the players wrote the answer on their own game boards. The second player stated that $3 \times 4 = 12$, and the leader said, "Correct, and Here I am!" All the players then wrote the product 12 on their game boards and marked that location with an X to indicate that they had found one of the message letters.

Subsequent players called $4 \times 4 = 16$, $7 \times 3 = 21$, and $3 \times 3 = 9$, and each time the leader responded, "Correct." Because the game is not finished, it will continue until the players place Xs in locations where seven of their answers match the message letters, and one or more players have been able to identify the answers to the multiplication fact problems and the correct letter locations. Any player to do so will be a winner!

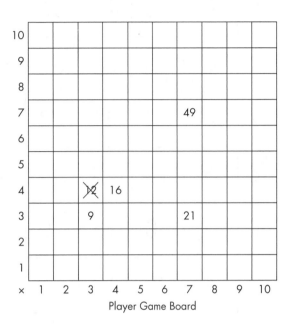

10	10	20	30	40	50	60	70	80	90	100
9	9	18	27	36	45	54	63	72	81	90
8	8	16	24	32	40	48	56	64	72	80
7	7	14	(H)	28	35	42	49	56	63	70
6	6	12	(E)	24	30	36	42	48	54	60
5	5	10	(R)	20	25	30	35	40	45	50
4	4	8	(E)	16	(I)	24	28	32	36	40
3	3	6	9	12	15	18	21	24	27	30
2	2	4	6	8	10	12	(A)	16	18	20
1	1	2	3	4	5	6	7	(M)	9	10
×	1	2	3	4	5	6	7	8	9	10

Master Game Board (kept hidden)

Player Game Board

Extensions:

1. At the outset, it is a good idea to use a large Player Game Board (on the chalkboard or overhead projector) to record the players' answers. This will reinforce correct multiplication products and also help clarify the written placement of the answers. (*Note:* The players are also informally learning about coordinate geometry.)

2. Change HERE I AM to some other short message. If, for example, it is Dan's birthday, the message might be DAY FOR DAN. In another instance, the players might be told that they will spell the name of the smallest state (RHODE ISLAND).

3. Students can logically analyze possible answer locations. In the Example shown above, a hit was made at $3 \times 4 = 12$, but not at 4×4 or 3×3. Discuss with students what other locations might possibly contain a HERE I AM letter, given that, as noted above, words can be placed horizontally, vertically, or diagonally, and that there will be a single empty space between words.

Master Game Board

10	10	20	30	40	50	60	70	80	90	100
9	9	18	27	36	45	54	63	72	81	90
8	8	16	24	32	40	48	56	64	72	80
7	7	14	21	28	35	42	49	56	63	70
6	6	12	18	24	30	36	42	48	54	60
5	5	10	15	20	25	30	35	40	45	50
4	4	8	12	16	20	24	28	32	36	40
3	3	6	9	12	15	18	21	24	27	30
2	2	4	6	8	10	12	14	16	18	20
1	1	2	3	4	5	6	7	8	9	10
×	1	2	3	4	5	6	7	8	9	10

Computation Connections

Player Game Boards

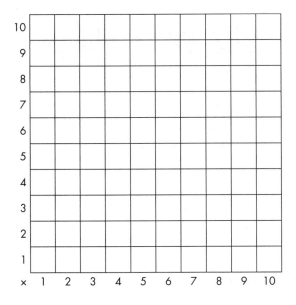

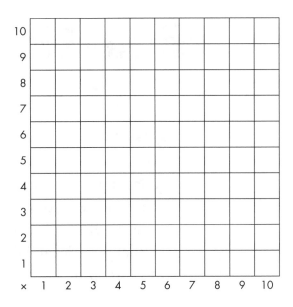

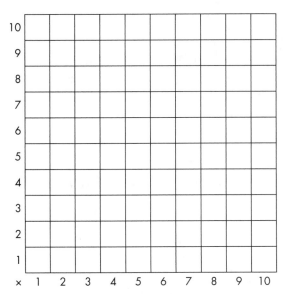

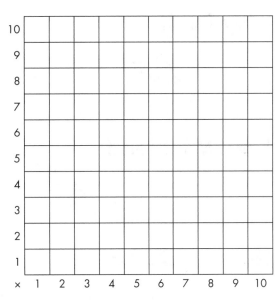

Chapter 52

Equation Match-Up

Grades 3–8

☒ Total group activity
☒ Cooperative activity
☐ Independent activity
☐ Concrete/manipulative activity
☐ Visual/pictorial activity
☒ Abstract procedure

Why Do It:

This activity provides either mental-math or pencil-and-paper practice with basic facts, computation, and problem-solving situations.

You Will Need:

Required are a supply of paper and place markers such as plastic discs, buttons, or beans.

How To Do It:

1. Each player must prepare an *Equation Match-Up* playing card by dividing a sheet of paper into a designated number of square areas, such as those shown in the Example. Discuss a series of number facts, computations, or problem situations with students, while they record the equations for them randomly on their card, plus leaving a "free" space anywhere on the card. At the same time, prepare matching answer cards, plus a few that do not match any of the equations discussed.

2. To begin play, mix the answer cards and place them face down in a pile. The top card is turned over, and if a player believes he or she has a match, a marker

is placed on top of that equation on the player's corresponding playing card. The first player or players to fill a line vertically, horizontally, or diagonally wins; however, he or she must verify each result by calling out the equations and their proper answers.

Example:

Notice that the match-up cards for the two players below contain the same equations, but in different positions.

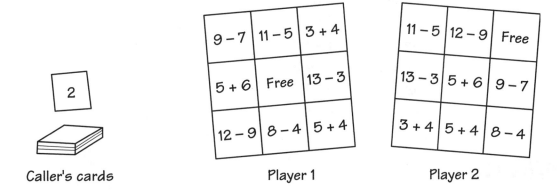

Caller's cards   Player 1   Player 2

Extensions:

1. Young players might begin by coloring dots to match numerals in each of the areas on their *Equation Match-Up* game cards. In order to play, they would place a marker on each dot diagram that matches an answer numeral.

2. *Equation Match-Up* can also be played in much the same manner as a *Bingo* game, in which columns are each designated by a letter; this version requires a pile of cards with the letters B-I-N-G-O written across the top. A letter card and an answer card are to be turned over at the same time, and the player must have an equation to match in the designated letter column. For example, an N and a 3 would mean that the player must have an equation with the answer of 3 under the N column.

3. Use several types of match-ups such as fractions and pizzas (for example, the numeral 1/3 and a picture or drawing of 1/3 of a pizza); geometric vocabulary and drawings (the word *pentagon* and a drawing of a pentagon); or simple word problems and solutions.

Chapter 53

Block Four

Grades 3–8
- ☐ Total group activity
- ☒ Cooperative activity
- ☐ Independent activity
- ☐ Concrete/manipulative
- ☐ Visual/pictorial activity
- ☒ Abstract procedure

Why Do It:

Students will practice with operations involving whole numbers, fractions, integers, or algebraic expressions.

You Will Need:

This activity requires a copy of a game board and "Record Board" (see reproducibles provided), two large paper clips, and pencils and scratch paper.

How To Do It:

In the activity *Block Four*, students will be practicing operations, marking their game board, and trying to win.

1. Put students into pairs. Give each pair a game board and a Record Board, and two large paper clips. There are three different types of game boards and a Record Board that can be photocopied. With the game boards provided, students can either work on multiplication facts with whole numbers, multiplying fractions, or adding fractions. Other types of game boards are mentioned in the Extensions.

2. Explain the rules of the game by putting the game board on an overhead and demonstrate playing against the class. Once the students understand the game, they can play with each other.

3. *Block Four* is played as follows. Player 1 places the two paper clips on two numbers, picked at random, at the bottom of the board. Player 1 will then multiply or add, depending on which game board they are using. Player 1 records his or her problem and answer on the Record Board. This player then puts an X on the correct answer on the board. Now it is Player 2's turn. Player 2 can only move one paper clip off of a number and place it on another number. The other paper clip stays where it is. Player 2 then performs the operation on the numbers, records it on his or her Record Board, and places an O on the correct answer on the game board. It is now Player 1's turn again. Player 1 can only move one paper clip again. The game continues in this manner until one player has won. To win the game, a player has to have four squares marked in a block, as shown below.

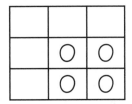

(*Note:* Players can place the paper clips on top of each other and then multiply or add the same two numbers together. Also, if multiple squares contain the same number, the player only chooses one on which to make his or her mark.)

Example:

In the following figure, Franco and Aaron are multiplying fractions and playing *Block Four*.

Extensions:

Extend this game by making other game boards. For example, a game board can be designed for students to practice adding or multiplying signed numbers. Also, beginning algebra students might use a board for multiplying binomials. Below are two possible variations on game boards. Only a portion of the board is shown.

Multiplying Integers

8	–12	16
6	–8	9
12	–6	–16

2 -3 3 **4** -4

Multiplying Binomials

$x^2 + 3x + 2$	$x^2 + x - 2$	$x^2 + 4x + 3$
$x^2 + 5x + 6$	$x^2 + 2x - 3$	$x^2 - x - 2$
$x^2 - 1$	$x^2 - 3x + 2$	$x^2 - 4$

$(x + 1)$	$(x + 2)$	$(x + 3)$	$(x - 1)$	$(x - 2)$
A	**B**	**C**	**D**	**E**

BLOCK FOUR
Multiplication Facts

1	2	3	4	5	6
7	8	9	10	12	14
15	16	18	20	21	24
25	27	28	30	32	35
36	40	42	45	48	49
54	56	63	64	72	81

1 2 3 4 5 6 7 8 9

$\dfrac{1}{9}$	$\dfrac{2}{9}$	$\dfrac{21}{64}$	$\dfrac{1}{36}$	$\dfrac{1}{4}$	$\dfrac{5}{32}$	$\dfrac{5}{72}$	$\dfrac{1}{54}$
$\dfrac{1}{8}$	$\dfrac{25}{48}$	$\dfrac{1}{32}$	$\dfrac{3}{16}$	$\dfrac{1}{36}$	$\dfrac{7}{24}$	$\dfrac{1}{6}$	$\dfrac{7}{72}$
$\dfrac{1}{12}$	$\dfrac{1}{2}$	$\dfrac{3}{32}$	$\dfrac{1}{9}$	$\dfrac{5}{24}$	$\dfrac{1}{16}$	$\dfrac{7}{32}$	$\dfrac{1}{16}$
$\dfrac{5}{16}$	$\dfrac{7}{48}$	$\dfrac{1}{24}$	$\dfrac{5}{48}$	$\dfrac{3}{32}$	$\dfrac{15}{64}$	$\dfrac{1}{6}$	$\dfrac{4}{9}$
$\dfrac{9}{64}$	$\dfrac{1}{8}$	$\dfrac{2}{27}$	$\dfrac{7}{12}$	$\dfrac{1}{3}$	$\dfrac{1}{27}$	$\dfrac{1}{24}$	$\dfrac{7}{16}$
$\dfrac{15}{32}$	$\dfrac{5}{12}$	$\dfrac{9}{32}$	$\dfrac{7}{64}$	$\dfrac{1}{12}$	$\dfrac{21}{32}$	$\dfrac{1}{4}$	$\dfrac{1}{64}$
$\dfrac{1}{18}$	$\dfrac{9}{16}$	$\dfrac{25}{64}$	$\dfrac{3}{8}$	$\dfrac{1}{48}$	$\dfrac{49}{64}$	$\dfrac{3}{64}$	$\dfrac{3}{16}$
$\dfrac{1}{72}$	$\dfrac{5}{24}$	$\dfrac{1}{81}$	$\dfrac{5}{64}$	$\dfrac{1}{8}$	$\dfrac{1}{18}$	$\dfrac{35}{64}$	$\dfrac{1}{12}$

$\mathbf{\dfrac{1}{2}}$ $\mathbf{\dfrac{1}{3}}$ $\mathbf{\dfrac{2}{3}}$ $\mathbf{\dfrac{1}{4}}$ $\mathbf{\dfrac{3}{4}}$ $\mathbf{\dfrac{1}{8}}$ $\mathbf{\dfrac{3}{8}}$ $\mathbf{\dfrac{5}{8}}$ $\mathbf{\dfrac{7}{8}}$ $\mathbf{\dfrac{1}{6}}$ $\mathbf{\dfrac{1}{9}}$

BLOCK FOUR
Adding Fractions

1	$\dfrac{3}{8}$	$\dfrac{11}{8}$	$\dfrac{3}{4}$	$\dfrac{3}{2}$	$\dfrac{19}{12}$	1	$\dfrac{31}{24}$
$\dfrac{5}{12}$	$\dfrac{5}{4}$	$\dfrac{7}{6}$	$\dfrac{13}{24}$	$\dfrac{7}{8}$	$\dfrac{5}{3}$	$\dfrac{23}{24}$	$\dfrac{4}{3}$
$\dfrac{1}{2}$	$\dfrac{9}{8}$	1	$\dfrac{41}{24}$	$\dfrac{5}{8}$	$\dfrac{1}{2}$	$\dfrac{7}{8}$	$\dfrac{13}{12}$
$\dfrac{5}{6}$	$\dfrac{7}{8}$	$\dfrac{19}{24}$	$\dfrac{25}{24}$	$\dfrac{35}{24}$	$\dfrac{19}{24}$	$\dfrac{3}{4}$	1
$\dfrac{29}{24}$	$\dfrac{11}{24}$	$\dfrac{1}{4}$	$\dfrac{4}{3}$	$\dfrac{17}{12}$	$\dfrac{17}{24}$	$\dfrac{5}{6}$	$\dfrac{3}{2}$
$\dfrac{7}{24}$	$\dfrac{7}{8}$	$\dfrac{11}{8}$	$\dfrac{3}{2}$	$\dfrac{7}{6}$	1	$\dfrac{3}{4}$	$\dfrac{29}{24}$
$\dfrac{9}{8}$	$\dfrac{37}{24}$	$\dfrac{3}{4}$	$\dfrac{7}{12}$	$\dfrac{13}{8}$	$\dfrac{5}{4}$	$\dfrac{13}{12}$	$\dfrac{1}{3}$
1	$\dfrac{5}{8}$	$\dfrac{11}{12}$	$\dfrac{9}{8}$	$\dfrac{2}{3}$	$\dfrac{1}{2}$	$\dfrac{9}{8}$	$\dfrac{25}{24}$

$$\dfrac{1}{2} \quad \dfrac{1}{3} \quad \dfrac{2}{3} \quad \dfrac{1}{4} \quad \dfrac{3}{4} \quad \dfrac{1}{8} \quad \dfrac{3}{8} \quad \dfrac{5}{8} \quad \dfrac{7}{8} \quad \dfrac{1}{6} \quad \dfrac{5}{6}$$

BLOCK FOUR
Record Board

Player 1 Problem	X Answer	Player 2 Problem	0 Answer

Computation Connections

Chapter 54

Silent Math

Grades 4–8

☒ Total group activity
☒ Cooperative activity
☐ Independent activity
☐ Concrete/manipulative activity
☒ Visual/pictorial activity
☒ Abstract procedure

Why Do It:

This activity requires players to think logically, find patterns, construct related diagrams, and practice basic addition and multiplication facts.

You Will Need:

All that is needed is a chalkboard or whiteboard and chalk or whiteboard pens.

How To Do It:

In this activity, you play an important role in guiding students through some diagrams and allowing the students to fill in the missing pieces. No talking is allowed in this activity. Start by putting diagrams, like those shown in Examples A–T, on the chalkboard. When the diagram contains all of the needed symbols and numbers, you indicate this by nodding yes. If some of the information is missing, make a questioning gesture. Players who think they know how to complete the diagram raise their hands; point to someone to finish the diagram. If the player completes the diagram correctly, shake his or her hand. Play continues in this way until it is clear that everyone has a good idea of what is happening. The ''no talking'' rule is then removed, and the players discuss their strategies. During a follow-up session, each player might be expected to contribute one or two partial diagrams to be solved by the other players.

Examples and Extensions:

The examples A through T below are sequenced from easy to more complex. Place them on the board one at a time, and either nod yes or make a questioning motion and appoint players to attempt solutions. Items A, B, E, F, and O are completed examples that can be shown to players to help them understand how to proceed. (Solutions to the unfinished problems are provided below, but should be used only if the players are truly "stuck.")

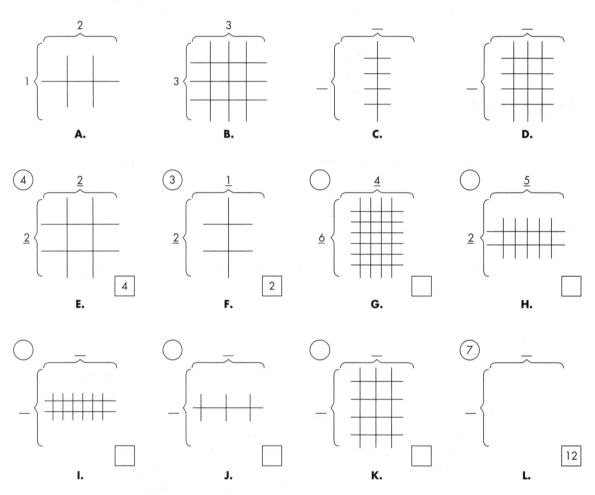

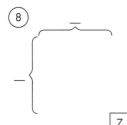 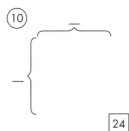 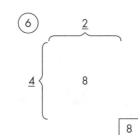 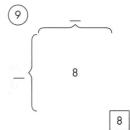

7	24	8	8
M.	**N.**	**O.**	**P.**

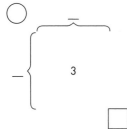 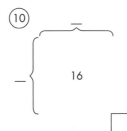 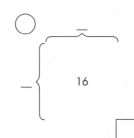 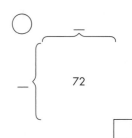

Q.	**R.**	**S.**	**T.**

Solutions:

C. 1, 4

D. 3, 4

G. + = 10, $x = 24$

H. + = 7, $x = 10$

I. 6, 2, + = 8, $x = 12$

J. 3, 1, + = 4, $x = 3$

K. 3, 4, + = 7, $x = 12$

L. 3, 4

M. 7, 1 ┼┼┼┼┼┼┼

N. 6, 4

P. 8, 1

Q. 3, 1, + = 4, $x = 3$

R. 8, 2, + = 10, $x = 16$

S. 4, 4, + = 8, $x = 16$;

or 8, 2, + = 10, $x = 16$;

or 16, 1, + = 17, $x = 16$

T. 8, 9, + = 17, $x = 72$;

6, 12, + = 18, $x = 72$;

or 4, 18, + = 22, $x = 72$;

or 2, 36, + = 38, $x = 72$;

or 1, 72, + = 73, $x = 72$

Chapter 55

Rapid Checking

Grades 4–8

☒ Total group activity

☒ Cooperative activity

☒ Independent activity

☐ Concrete/manipulative activity

☐ Visual/pictorial activity

☒ Abstract procedure

Why Do It:

Rapid Checking provides students with a quick and captivating way to check answers to addition, subtraction, multiplication, and division problems.

You Will Need:

A chalkboard or whiteboard on which to demonstrate the method is needed, and students will need pencils and paper.

How To Do It:

To quickly check any computation problem, students can add the digits together (repeatedly if needed) for each of the initial problem numbers until a single-digit "representative number" is reached. They then redo the problem process (add, subtract, multiply, or divide) with the representative numbers, and then add those digits to get a single-digit representative answer. Next, they go to their original answer and add those digits together until they get another single-digit representative answer. Finally, they compare their two representative answers; if they are the same, the answer checks.

In the figure below, the students are checking the answer to the problem 6,492 multiplied by 384. First, they added the digits in 6,492 and got 21, then they added the digits in 21 and got 3. They stopped there, because 3 is a single digit number. Next they added the digits in 384 and got 15, then added the digits in 15 and got 6. At this point they multiplied 3 and 6 to get 18, then added the digits of 18 and finally got 9. Last step was to add the digits in the answer 2,492,928 and they computed 36, then added the digits in 36 and got 9. This result is the same as the 9 they got previously; therefore their answer to the multiplication problem is correct. (*Note:* Please read the tips for checking subtraction and division in the Examples. Also, an error possibility is discussed in Extension 3.)

Examples:

1. Remind students to add digits to obtain single-digit representative numbers as they follow the rapid checking of the problem below.

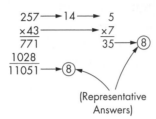

(Representative Answers)

2. Now have students try the same process with a column addition problem.

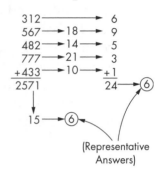

(Representative Answers)

3. When rapidly checking a division problem, students may benefit from thinking of the procedure in terms of multiplication. The quotient and divisor are multiplied together to get the dividend. Make sure students include any remainders in their answers.

Think:

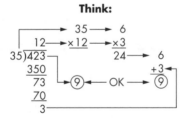

4. Students are most readily able to check subtraction computations, such as the one shown below, when they think of them in terms of addition. The difference and the number being subtracted are added to get the number being subtracted from.

Think:

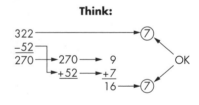

Extensions:

1. Have students see if the rapid checking works for relatively easy problems, such as $12 + 45$ or 8×9.

2. Students may use the method to check decimal problems, such as $0.97 + 0.42 + 0.38$ or $0.4321 + 0.5 + 0.892$.

3. Be careful that students do not switch the digits around in their original answers; if they should do so, the rapid check would falsely confirm their answers. For instance, in the addition problem shown in Example 2 above, although the true answer is 2,571, if one mixes the digits to read 2,517, the representative outcome would be 6 in either case. (Such errors happen infrequently. Thus most answers rapidly checked will be correct.)

Computation Connections

Section Three

Investigations and Problem Solving

Students cannot be prepared for every problem they will encounter throughout life. However, they can and should be exposed to a wide variety of situations warranting investigation, and should be equipped with problem-solving strategies. The activities in this section stem from a variety of real-life situations and include both written and verbal word problems; problem-solving plans; problems with multiple answers; and investigations that incorporate spatial thinking, statistics and probability, measurement, and scheduling. Because many of the tasks are hands-on and nearly all call for direct participation, students will be highly engaged and have fun with the learning process.

Activities from other parts of this book can be used to help young learners develop problem-solving skills. Some of these are *Everyday Things Numberbooks* (p. 7), *Celebrate 100 Days* (p. 27), and *A Million or More* (p. 62) from Section One, *Dot Paper Diagrams* (p. 112) and *Silent Math* (p. 203) in Section Two, and *Problem Puzzlers* (p. 392) and *String Triangle Geometry* (p. 411) from Section Four.

Chapter 56

Shoe Graphs

Grades K–3

☒ Total group activity
☒ Cooperative activity
☐ Independent activity
☒ Concrete/manipulative activity
☒ Visual/pictorial activity
☒ Abstract procedure

Why Do It:

Students will investigate everyday applications of mathematics and organize the information into statistical graphs.

You Will Need:

One shoe from each student is required, in addition to a yardstick, a marking pen, and masking tape.

How To Do It:

Use the masking tape to mark out a floor grid of 3 or 4 columns by 10 to 12 rows (the grid spaces should each measure about 1 square foot). Label each column according to the type of shoe that may be placed in it: (1) slip-ons, (2) tie shoes, (3) Velcro fasteners, and (4) other types. At the baseline, have each student place 1 of his or her shoes in a matching labeled column. Then initiate a discussion based on such questions as "How many people were wearing slip-on shoes? How many more people were wearing slip-ons than shoes with Velcro fasteners?" During this discussion, the yardstick may be used as a marker by placing it at the top of one shoe column (as in the following figure) such that the number of additional shoes of a select type can be easily viewed and counted. When students are ready, or for advanced students, continue the analysis with such a discussion as "We counted 8 tie shoes in that column. So if 8 people are wearing tie shoes, and each needs 2 shoes, how can we find out the total number of shoes those 8 people must have?" (*Note:* Most adults would say that $8 \times 2 = 16$, but many young students will not yet have acquired multiplication skills. Such beginners might be helped to count the occupied column spaces by 2s.)

Example:

In the situation shown above, the students have each placed one of their shoes in a column that matches. The teacher is now asking one of a series of questions that will help the learners analyze their real graph data.

Extensions:

Following students' initial experiences, in which they considered how many total shoes, how many more or less, and so on, learners might be asked to consider some of the following possibilities.

1. Investigate what color hair most people have. Have students stand in columns on the floor graph according to hair color (blond, brown, black ...). Ask such questions as "If there are 9 people with blond hair in this class, and there are 8 classes at this school, how many blond people might there be in the whole school?" Allow the students to work in small groups as they attempt to find an answer. When they think they have a solution, ask them to explain their thinking. Probe further and ask whether there is a single answer to the problem, or whether their solution is an estimate. (*Note:* Construct similar graphs for color of eyes, type or color of shirt being worn, and so on.)

2. Using personal photos of the students when building graphs is a very effective technique. When a supply of individual photos are available, various types of data can be considered, such as favorite flavors of ice cream, number of pets each family has, preferred physical education activity, or favorite thing to do on weekends. Once such data has been collected, the class can compile representative graphs by pasting the students' individual photos in the appropriate columns. Learners should be asked to analyze the data in as many ways as possible. (*Note:* Because the activity will require a number of pictures of each student, photocopy the class photograph to help reduce costs.)

Chapter 57

Sticky Gooey Cereal Probability

Grades K–3

☐ Total group activity

☒ Cooperative activity

☒ Independent activity

☒ Concrete/manipulative activity

☒ Visual/pictorial activity

☒ Abstract procedure

Why Do It:

This activity uses a simulation to solve a probability problem that students may experience in real life. This technique can be used to solve other problems that might be of interest to the students.

You Will Need:

One spinner (pattern provided) or die is required for each group or individual. Also necessary are copies of the "Sticky Gooey Cereal Record Sheet" (reproducible provided) and pencils. Also, graph paper and markers are needed if a graph is to be made.

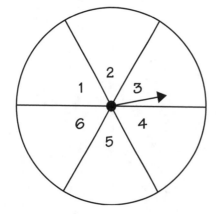

How To Do It:

1. Introduce learners to the following problem and have them try to answer the question.

> A cereal company has randomly placed 6 different prizes in boxes of Sticky Gooey Cereal, with only one in each box. All 6 prizes are evenly distributed. When you buy a box of cereal, you do not know what the prize will be. Do you think that you have a good chance of getting all 6 prizes if you buy 10 boxes of cereal?

2. Students should make spinners with six sections numbered 1, 2, 3, 4, 5, and 6. They could also draw a picture of a different prize in each section. (A die would work for this simulation.) To make a spinner, use the Cut-Out Spinner, and see the explanation for its use provided in the activity *Fairness at the County Fair* (p. 321).

3. Groups or individuals will spin the spinner 10 times (or toss the die 10 times), and after each spin they should record the number they got on the record sheet as part of Trial 1. If all six numbers on the spinner show up in the 10 spins, then the student will put an X under "YES" in the column "Did I Get All 6 Prizes?" If all six numbers do not show up in the 10 spins, the student will mark "NO." This process is repeated to complete six trials.

4. After the six trials, each group or individual will count how many marks they have in the "YES" column and how many they have in the "NO" column.

5. Finally, they can answer the initial question using their results, with the help of the Results Chart at the bottom of the Sticky Gooey Cereal Record Sheet.

Extensions:

1. Have students tally up all the "YES" and "NO" marks for the whole class and see if the answer to the question turns out different from individual outcomes.

2. Extend the problem to include 8 prizes and make a spinner with 8 equal sections. Students can again spin 10 times, or can extend their spinning to 15 or 20 times.

Sticky Gooey Cereal Probability

3. Apply this type of simulation to help answer other questions students might have. For example: A penny candy machine has 12 different types of candy in it. Assuming that there is a large number of candies equally divided among the different types, how many pennies will you have to use to get one of each type of candy? If a spinner or a die do not work well (for example, if the number of prizes in the Sticky Gooey Cereal activity was 20), you can also pull pieces of paper out of a paper bag, replacing them each time.

Cut-Out Spinner

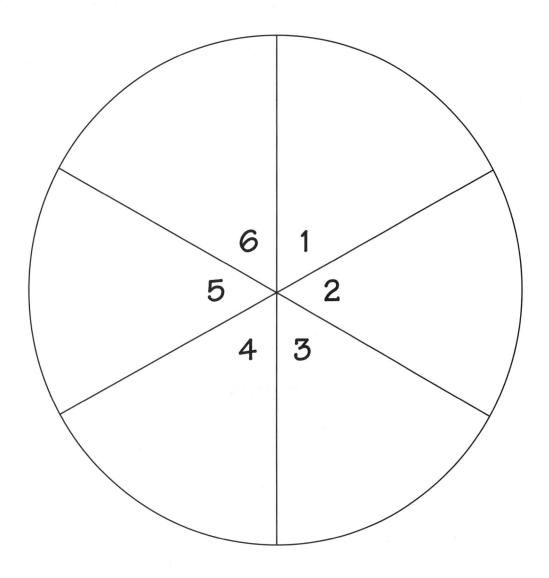

Sticky Gooey Cereal Record Sheet

Trial	Result	Did I Get All 6 Prizes? "YES"	"NO"
Example	1, 4, 6, 2, 1, 1, 3, 4, 6, 6		X
1			
2			
3			
4			
5			
6			
How many are marked "YES" and how many are marked "NO"? (Do not count the example.)			

RESULTS CHART

Number of Marks in "YES" Column	Answer to Question
5 or 6	Yes, I have a very good chance.
3 or 4	Probably not, so I should buy more.
1 or 2	No, I do not have a good chance.

Chapter 58

Sugar Cube Buildings

Grades 1–8

- ☒ Total group activity
- ☒ Cooperative activity
- ☒ Independent activity
- ☒ Concrete/manipulative activity
- ☒ Visual/pictorial activity
- ☒ Abstract procedure

Why Do It:

Students will use logical-thinking skills in an applied geometry problem.

You Will Need:

Initially each student will need 4 sugar cubes; at more advanced stages, each group, or each individual student, may wish to work with as many as 10 sugar cubes at a time. Other items can be used, such as connecting blocks that snap or pop together. For the most advanced investigations, each group may need to work with 36, 48, or 100 sugar (or other) cubes. Students will need graph paper to keep records of their findings (see *Number Cutouts*, p. 22, for graph paper that can be photocopied), and those completing further investigations might also need "3-D drawing paper" (see reproducible page) and pencils.

How To Do It:

1. When first introducing *Sugar Cube Buildings*, give students 4 sugar cubes each and tell them that they need to accomplish several "jobs." Their first job is to design and build as many single-story buildings as

possible, with the requirement that each sugar cube be fully attached to the side of at least one other cube (see functional design and unacceptable design in figures below). As they find workable arrangements, students should record the designs as top views on graph paper, as shown in the figures below. When finished, the students should be allowed time to discuss, compare, and contrast their findings. If desired, it is easy to construct permanent models of the building designs by moistening selected surfaces of the sugar cubes, placing them tightly together in an approved design, and allowing them to dry.

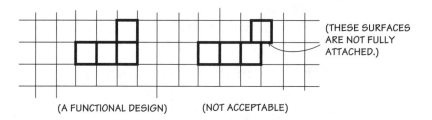

(THESE SURFACES ARE NOT FULLY ATTACHED.)

(A FUNCTIONAL DESIGN)   (NOT ACCEPTABLE)

2. The student's second job, again using 4 sugar cubes, is to design as many multiple-level buildings as possible, with the additional requirement that all cubes, except those on ground level, must be fully supported (see figures below). Students should record their findings as side views, using either graph paper or the 3-D drawing paper.

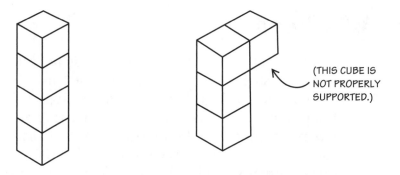

(THIS CUBE IS NOT PROPERLY SUPPORTED.)

3. A third job involves giving the students more cubes, perhaps 8, and asking them to design all possible buildings of 1 story, 2 stories, 3 stories, or up to 8 stories. A few of the possible designs are shown on 3-D drawing paper in the Example.

Example:

Shown here are some of the student designs for this activity, created while working with 8 sugar cubes.

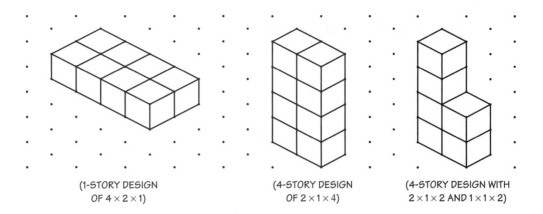

(1-STORY DESIGN
OF 4 × 2 × 1)

(4-STORY DESIGN
OF 2 × 1 × 4)

(4-STORY DESIGN WITH
2 × 1 × 2 AND 1 × 1 × 2)

Extensions:

1. Challenge the students to use a large number of sugar cubes (perhaps 36, 48, or 100) for their next job. With the specified number of cubes, they are to create all possible buildings that are rectangular solids (they might think of these as solid box shapes similar to the first two figures in the Example above). Students should record, discuss, compare, and contrast their findings. In particular, they should note any patterns they discover. For example, for 36 cubes, the 6 by 6 rectangular solid is the only design that is a large cube, and that is because 36 is a perfect square number; 48 sugar cubes will not yield a rectangular solid that is a large cube.

2. For advanced classes, assign costs per square unit and have students determine the total price for different building designs that use the same number of sugar cubes. For example, have students use the prices listed below to determine the costs for all the possible different 4-cube buildings. They may also determine the cost for buildings that use more, or fewer, sugar cubes.

Costs: Roof = $5,000 per square unit
Floor (or land) = $10,000 per square unit
Outside Walls = $3,000 per square unit

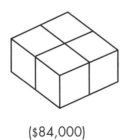

($84,000)

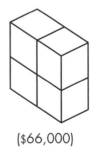

($66,000)

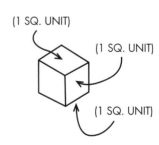

(1 SQ. UNIT)

(1 SQ. UNIT)

(1 SQ. UNIT)

3-D Drawing Paper

Investigations and Problem Solving

Chapter 59

A Chocolate Chip Hunt

Grades 1–8

☒ Total group activity

☒ Cooperative activity

☒ Independent activity

☒ Concrete/manipulative activity

☒ Visual/pictorial activity

☒ Abstract procedure

Why Do It:

Students will use estimation, data gathering, information organization, and logical-thinking skills while investigating a real-life application of mathematics.

You Will Need:

Several packages of commercially baked chocolate chip cookies are needed so that each student will get at least three cookies. Napkins, pencils, and a "Chocolate Chip Records" page (one for each student) also are needed. (*Note:* This activity should not be used for classes in which some students have allergy or other eating restrictions.)

How To Do It:

The goal of this activity is for students to experience collecting actual data and to analyze that data. This is the main premise behind the study of statistics.

1. Begin by asking students who would like to have a chocolate chip cookie, a question to which most will respond positively. Tell them that they need to wash their hands because their next math activity uses chocolate chip cookies and, if they wish, they will be able to eat the cookies when finished.

2. Pass out a "Chocolate Chip Records" page, napkins, and one cookie (use the same brand for everyone) to each participant. The students' first math job is to estimate how many chocolate chips they think are in their cookies and to write their estimates on the records pages. Their next job is to break their individual cookies into small pieces in order to locate and count all of the chocolate chips, recording the numbers they find. (If students wish, they may now eat their cookies.) At this point, the teacher, or a leader, should begin to ask for and organize the individual findings on the chalkboard (or on butcher paper or an overhead transparency). After the class has discussed these data, students should cooperatively develop a bar graph portraying the information they gathered.

3. Next ask the students, "If I give you another cookie of the same brand, how many chocolate chips will this new cookie have? Will it have the same number of chips as your first cookie? On your recording sheet, write down your estimates and tell why you think as you do." Give students a second cookie with which to complete this new math job by counting the actual number of chocolate chips. Again seek and record the chocolate chip information and, with students' help, organize the data on the chalkboard. Students should also develop another bar graph to compare and contrast with that from the first trial.

4. A third math job involves different brands of chocolate chip cookies. Using these, the learners should estimate the number of chocolate chips per cookie and record their predictions. Students then break up the cookies to count the actual chips; record, organize, and graph this information; and compare, contrast, and discuss their findings. They may also, if they wish, eat the data! Their comparisons of the numbers of chocolate chips in the different brands might lead to a discussion of which brand they would prefer to buy and what other factors (such as expense) might determine their choice.

Example:

In the situation pictured below, the students are comparing two different brands of cookies to find out which is the "chocolate chippiest"!

Extensions:

1. The activities for young learners might include estimating, counting, tallying, and graphing the chocolate chips, as well as comparing and contrasting their findings. Middle grade (grades 4–6) and older students should also compare brands based on their price versus their value.

2. Able students should also compare the chocolate chip cookie data in terms of means, medians, and modes. For instance, when assessing the number of chips in one brand of cookies, the learners might find the chocolate chip range (the highest minus the lowest number of chips), the chip mean (the average computed by adding all the number of chips and dividing by the number of cookies), the chip median (the "middle" number when chips are arranged from lowest to highest), and the chip mode (the number found most frequently). Mean, median, and mode are explained further in *The Three M's* (p. 290).

A Chocolate Chip Hunt

Chocolate Chip Records

What I found when checking for chocolate chips in one brand of cookies was:

	1st Cookie	2nd Cookie	3rd Cookie
Estimated Number of Chips			
Actual Number of Chips			

What the whole group found when checking 1 brand of cookies for chips was:

Class Chocolate Chip Graph for _____ **Brand of Cookies**

People Who Found This Number of Chocolate Chips		
10		
9		
8		
7		
6		
5		
4		
3		
2		
1		

Number of Chocolate Chips per Cookie

Chocolate Chip Records (continued)

What we found when checking several brands of cookies for chocolate chips was:

Class Chocolate Chip Graph for Several Brands of Cookies

Brands
of
Cookies

Number of Chocolate Chips per Cookie

Chapter 60

Flexagon Creations

Grades 1–8

☒ Total group activity
☒ Cooperative activity
☒ Independent activity
☒ Concrete/manipulative activity
☒ Visual/pictorial activity
☒ Abstract procedure

Why Do It:

Students will analyze and understand the attributes of geometric 3-dimensional figures.

You Will Need:

This activity requires enlarged flexagon patterns (see page 230), a supply of card stock (discarded file folders work well), rubber bands, pencils, scissors, and the "Can You Create a Flexagon?" handout. A math dictionary or math textbooks with good glossaries may also prove helpful.

How To Do It:

In this activity, students will create polyhedrons (3-dimensional figures with faces that are polygons) with the help of flexagons. A flexagon is a polygon with flaps that can be folded and connected to other flexagons to create the polyhedron. Have the students trace enlarged triangle and square flexagon patterns on card stock, cut them out, and fold the tabs in both directions. Approximately 10 of each pattern will be needed for each group or individual. Next, provide rubber bands and allow the students to explore what happens when the tabs of the flexagons are banded to each other. They will soon see, for example, that 4 triangles rubber-banded together make a closed figure (a triangular-based pyramid), and 6 squares banded together can make a cube. Now distribute the "Can You Create a Flexagon?" handout and have students predict whether they will be able to build the suggested geometric configurations. They should then attempt to construct closed 3-dimensional figures using flexagons and rubber bands. To connect two sides of the 3-dimensional figure, fold the flaps on the flexagons, and after placing two flaps from two different flexagons together, put the rubber around the flaps to keep them together. Discuss with the students whether they were able to make a closed figure with the flexagons mentioned in the chart and why. The students might also wish to know the names for each figure; various reference sources may help, but a mathematics dictionary or math textbooks containing good glossaries will be the most helpful.

Example:

Shown below are some student-designed flexagons. Student 1 has used a square and 4 triangles to create a square-based pyramid, and Student 2 has utilized 10 squares to build the framework for a rectangular solid.

Extensions:

1. Challenge the learners to create geometric shapes other than those they built when completing the "Can You Create a Flexagon?" handout. They may use as many triangle and square flexagon pieces as they wish.

2. Ask students to construct flexagon patterns with 5, 6, 7, 8, or more edges, and then to build "new" flexagons and keep a record of the number of faces of each type, the number of edges, the number of vertices, and the figures' names.

3. Students can build slightly different frameworks with plastic straws (of equal lengths) and paper clips. To do so, students must insert a paper clip into each end of a straw, leaving loops of about 1/8 inch extending. They then hook another paper clip into each loop, insert this second paper clip into another straw, and continue this process until they have created the desired polyhedron.

4. A neat extension, once the plastic straw and paper clip frameworks have been built (see Extension 3 above), is to dip these structures in soap film and have the class record the bubble configurations that result. The students, and maybe the teacher too, will be surprised. (*Note:* The soap film is of the same type used in children's bubble-blowing sets. You may make your own with 2/3 cup dishwashing liquid, 1 gallon of water, and 2 to 3 tablespoons of glycerin [available in pharmacies or chemical supply stores]. The glycerin is not necessary, but should make the bubbles stronger. A Web site containing more bubble formulas is http://bubbleblowers.com/homemade.html.)

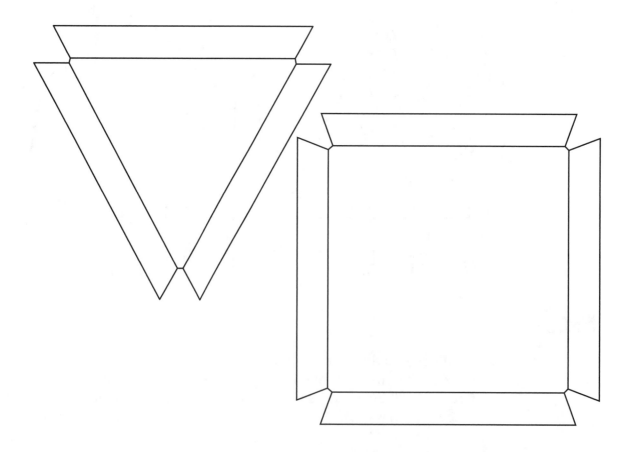

Investigations and Problem Solving

Can You Create a Flexagon?

Can You Make a Closed Figure With	Prediction (Yes or No)	Build It If You Can. Were You Able to?	What is the Mathematical Name for This Flexagon?
3 triangles?			
4 triangles?			
5 triangles?			
6 triangles?			
3 squares and 1 triangle?			
3 squares and 2 triangles?			
5 squares and 4 triangles?			
6 squares and 2 triangles?			
1 square and 4 triangles?			
1 square and 5 triangles?			
4 squares?			
6 squares?			
8 squares?			
10 squares?			

Use triangles and/or squares and try to create some more flexagons. In the spaces below, list any that you found.

What did this investigation show? Describe it with a simple statement.

Chapter 61

Watermelon Math

Grades K–8

☒ Total group activity

☒ Cooperative activity

☐ Independent activity

☒ Concrete/manipulative activity

☒ Visual/pictorial activity

☒ Abstract procedure

Why Do It:

This activity gives students hands-on experience with estimation, counting, place value, computation, and graphing through working with a familiar food.

You Will Need:

Watermelon Math requires a watermelon (or another fruit or vegetable with a lot of seeds); string; scissors; 1-, 5-, and 10-pound weights (or use 1 pound of butter, 5 pounds of sugar, and a 10-pound sack of flour); a weight scale; graph paper; plastic or paper cups; napkins; a large knife; and pencils and paper.

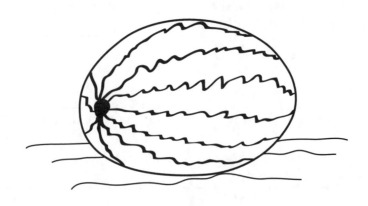

How To Do It:

1. Secretly bring the watermelon to class in a box. Students should try to find out about the contents of the box by asking attribute questions that can be answered yes or no. (They may ask such questions as ''Is it spherical [or round shaped],'' but not ''Is it a soccer ball?'') When they are quite certain they know what it is, a last ''Is it a (specific item)?'' question is allowed. Then hold the watermelon up for everyone to see.

2. Continue by providing each participant with a long piece of string. Ask students, ''How big around (the girth) is the watermelon?'' Then have each person cut off a piece of string that they estimate to be the proper length. Wrap a new piece of string around the watermelon and cut it off to equal exactly the girth. Then make a string graph with the categories Too Short, Just Right, and Too Long by suspending all of the strings from the top of a bulletin board in ascending order. Discuss the students' findings and how they might make closer estimates in the future.

3. Next the learners explore the weight of the watermelon, for example by lifting it and comparing it with pound (or metric) weights (or with 1 pound of butter, 5 pounds of sugar, and a 10-pound sack of flour). After students have estimated and recorded how many pounds they think the watermelon weighs, the melon should be weighed using a scale. Then, as a class, develop and discuss a graph indicating Too Light, Just Right, and Too Heavy based on students' estimates.

4. Ask the participants to wash up because next they'll get to eat the watermelon (excluding the seeds). (*Note:* If any students have an allergy or other eating restriction for watermelon, a different fruit or vegetable with seeds could be used.) Before eating, however, students must estimate how many seeds will be in the watermelon. After students have helped decide how to divide the melon into pieces that are about equal, tell them what fraction each piece will be, cut the watermelon into pieces, and give these to the class to eat. When students have finished eating, provide small cups (clear plastic are best) and have them place 10 seeds in each cup (it might be helpful to have them wash and dry the seeds on a paper towel first). Together, count all of the seeds by 10s (10, 20, 30 . . .) to reach the total. (*Note:* Pile 10 cups together to equal 100s. When doing so, be certain to help students make the place value connections.) Furthermore, another graph may be constructed to note and discuss Too Few, Just Right, and Too Many in regard to students' seed estimates.

5. A further activity to determine how much of the watermelon was rind and how much was edible might be useful. This can be done, of course, by weighing all of the leftover rind pieces together and subtracting this value from the total weight of the melon, which had been determined earlier.

Example:

In the illustration below, the learners are counting the watermelon seeds, putting 10 in each clear plastic cup, and stacking 10 cups together to make 100s.

Extension:

Use other fruits and vegetables for similar investigations. Apples or oranges work especially well when students are working individually. Pumpkins or squash are also suitable for large group investigations. Even peanuts (beware of allergies!) can be investigated as to the number of nuts in each shell, how many it takes to equal a pound, how much space they take up inside versus out of the shells, and so on.

Chapter 62

Restaurant Menu Math

Grades K–8

☒ Total group activity

☒ Cooperative activity

☒ Independent activity

☐ Concrete/manipulative activity

☒ Visual/pictorial activity

☒ Abstract procedure

Why Do It:

Students will practice computation with whole numbers, decimals, and percents in a real-life situation.

You Will Need:

A supply of restaurant menus are required (if asked, many restaurants will provide old menus for free; or menus may be photocopied), plus pencils and paper.

How To Do It:

Have each student decide with whom they might be going to breakfast on Sunday morning. It then becomes their job to ''take the orders'' from the classmates they are eating with and determine the total cost for their breakfasts, including tax and tip. Students could also decide to choose their family as breakfast mates and write down what they think their family members might order. Each student should have a list of items from the menu and their prices for themselves and anyone else they choose to have breakfast with. The students can also check one another's bills for accuracy or work with money (real or facsimile) to make change for other classmates.

Example:

Shown on page 237 is the breakfast portion of a menu. A student and his parents will be eating breakfast together. The students are to determine which items each family member will order and calculate what the total cost will be.

Extensions:

1. The menus for young children might be made easier by pricing all items as $1, $2, $3, and so on. (To do this, place blank stickers over the original prices and write in the nearest even dollar amount.)

2. Students can compare meals from different restaurants. Have them select a typical meal from a variety of different restaurant menus and then discuss such factors as quantity, variety (for example, whether a salad is included), cost, and even restaurant atmosphere. This could be a week-long project that students do for homework, bringing their results to class, or students could collect different menus from a variety of restaurants and bring them into class for students to use. Students can also determine what the same meal would cost to prepare at home.

3. Some menus have the calorie count next to each item, or have a diet portion of the menu with a calorie count. There are books available for purchase, as well as Web sites (for example, www.sparkpeople .com), that have the approximate calories for various items of food. Older students can discuss the calorie (or sugar, fat, or protein) content of certain foods, find a meal on a menu, and compute the total calories (or sugar, fat, or protein) for that meal. This could even be integrated with a science lesson on what the human body does and does not need.

Try our Fresh Squeezed Orange Juice!

By the Glass

Large	2.40	Orange Juice
Regular	1.95	Apple Juice
Liter	5.95	Tomato Juice
		Cranberry Juice

4-Egg Omelets

Served w/your choice of hash browns or home fries, & toast or 2 small pancakes

California Omelet 7.95
Filled w/bacon, tomatoes, mushrooms & Cheddar cheese, topped w/avocado slices.

Ham & Cheese Omelet 7.75
Chopped ham & shredded Cheddar cheese.

Veggie Omelet 7.95
Onions, mushrooms, bell peppers & Cheddar cheese, topped w/avocado slices & sour cream.

Meat Lover's Omelet 7.95
Ham, bacon, sausage & onion w/Swiss & Cheddar cheese.

Party Omelet 7.95
Ham, bacon, bell peppers, mushrooms, onions & tomatoes topped w/Cheddar cheese.

New Taco Omelet 7.65
Taco meat & cheese, topped w/salsa & sour cream.

New Chicken Enchilada Omelet 7.95
Diced chicken, onions, bell pepper and cheese, topped w/enchilada sauce, cheese & chives.

Breakfast Skillets

Skillets are served over a mound of seasoned home fries & topped w/2 eggs, served w/toast or 2 small pancakes.

Meat Lover's Skillet 7.95
A Perfect blend of chopped ham, bacon & sausage w/mushrooms, onions & tomatoes, topped w/shredded Cheddar cheese.

Scrambler Skillet 7.45
2 scrambled eggs w/mushrooms, chives & sliced avocado over a mound of seasoned home fries, topped w/Cheddar cheese & served w/country gravy & toast.

Steak Ranchero Skillet 7.95
Seasoned New York steak chunks sauteed w/onions, bell peppers, mushrooms, tomatoes & cilantro, served w/toast or a flour tortilla & salsa on the side.

A La Carte

Bacon or Sausage	3.75
Ham Slice	3.95
Hash Browns or Home Fries	2.95
Bagel & Cream Cheese	2.85
2 Biscuits & Gravy	3.65
1 Egg	.95
2 Egg	1.95
Toast	1.55
English Muffin	1.65
Oatmeal	2.95
Fresh Fruit (Bowl)	4.65
Fresh Fruit (Cup)	2.95

Visit us Online at
www.JacksRestaurantRocks.com

Delicious Coffees

Coffee - Farmer Bros. Premium	1.60
2 free refill.	
Cafe Au Lait	3.25
Cappuccino	3.25
Cafe Mocha	3.45
Italian Soda	2.25
Raspberry, Cherry, Strawberry, or Vanilla.	
Snicker Doodle	3.55
Flavor Shots	.60
Vanilla, Hazelnut or Caramel.	
Extra shot of Espresso	.60
Mexican Mocha	3.45

BREAKFAST SERVED ALL DAY & ALL NIGHT

Pancakes and Stuff

Full Stack of Golden Pancakes (3)	4.75
Short Stack (2)	4.45
Belgian Waffle	5.25
Texas-Style French Toast (2 Slices)	4.45
Texas-Style French Toast (3 Slices)	5.95
French Toast Special	5.95
2 Slices of French Toast w/2 eggs & sausage links or 2 bacon strips.	
Belgian Waffle Combo	6.45
Belgian Waffle w/2 eggs & 2 strips of bacon or 2 sausage links.	

Strawberries

Waffles served 23 hours

Served Over Your Choice:
Pancakes,
Waffles or French Toast
Top it off with whip cream!
6.45

Biscuit & Gravy 5.25

Served with w/sausage patty.

Big Boy's Breakfast Specialties

The Lumberjack 9.65
3 eggs, 3 sausage & hash browns or home fries, served w/6 pancakes & toast.

Breakfast Burrito 7.45
2 scrambled eggs, sausage, home fries & cheese, wrapped in a flour tortilla & covered w/more cheese. Served w/salsa on the side.

Eggs and Things

Served w/hash browns or home fries & toast or 2 small pancakes.

Linguisa & Eggs	7.45
Bacon or Sausage & Eggs	6.95
4 strips of bacon or 3 sausage links	
Ham & Eggs (6 oz. pit ham)	7.35
Corned Beef Hash & Eggs	6.95
2 Eggs (any style)	4.95
Diced Ham Scrambled	6.65
Mini Breakfast	3.45
Slice of French Toast & 2 Eggs	

Jack's Steak & Eggs Specials

Served w/hash browns or home fries & toast or 2 small pancakes.

1/2 lb Harris Steak & Eggs
Served w/2 eggs, hash browns or home fries. 7.95

N.Y. Steak (1/2lb) & Eggs
Served w/2 eggs, hash browns or home fries. 7.95

1 lb N.Y. Steak & Eggs
Served w/3 eggs, hash browns or home fries & biscuit & gravy. 10.45

16 oz. Chicken Fried Steak & Eggs
Served w/3 eggs, hash browns or home fries and toast. 10.45

Jack's Country Breakfast	7.45
2 eggs, 1 slice of ham & 2 pancakes, served w/hash browns or home fries.	
Jack's Sampler	8.25
2 eggs, 2 small slices of ham, w/bacon & 2 sausage, served w/hash browns or home fries & a biscuit w/country gravy.	
Sleepy Dan	8.25
2 eggs, 3 bacon, 3 sausage, served w/hash browns or home fries, plus 3 pancakes & toast.	

Beverages

Hot Tea - regular or spiced	1.60
Iced Tea or Iced Coffee	1.95
Soft Drink (20 oz.)	1.95
One free refill.	
Pink Lemonade	1.95
One free refill.	
Regular Milk (Small)	1.75
Regular Milk (Large)	2.15
Hot Chocolate	1.95
One free refill.	

Chapter 63

Peek Box Probability

Grades K–8

☒ Total group activity

☒ Cooperative activity

☒ Independent activity

☒ Concrete/manipulative activity

☒ Visual/pictorial activity

☒ Abstract procedure

Why Do It:

Students will gain experience with sampling techniques for the purposes of collecting, organizing, and interpreting data.

You Will Need:

A small box is required for each group or individual (checkbook boxes work well), as well as at least 10 marbles (2 different colors) per box and copies of the "Peek Box Records" worksheet. This activity could also be done with paper bags and poker chips, or anything that has the same shape but different colors.

How To Do It:

This activity helps students discover an important fact in the study of statistics: the more frequently they perform an experiment, the more likely it will be that their conclusion is accurate.

1. Using at least 10 marbles per box, secretly place the marbles in the boxes in selected color rations (for example, 6 red and 4 blue). For the first trial, all boxes should contain the same ratio of marbles. Cut a corner off of the boxes small enough to prevent the marbles

from coming out. Also pass out copies of the "Peek Box Records" worksheet so that students may tally and analyze their findings.

2. Tell the students that each of their boxes contains 10 marbles, some of which are red and some blue. (*Note:* Do not tell the students that each box contains the same ratio of marbles.) It is their job to shake the boxes 10 times each and keep a record, on their "Peek Box Records" worksheet, of how many reds and how many blues they see through the hole in the box. When finished, discuss what their tallies were and what numbers of red and blue marbles they think are in the Peek Boxes and why. Then open the boxes to let them see that all ratios were the same, and allow further discussion. Repeat the process several times with different ratios of marbles in the boxes and with a discussion of students' predictions. (If using poker chips and paper bags instead of marbles and boxes, have students first draw a chip out of the bag after each shake without looking in the bag, and then replace the chip in the bag before the next shake.)

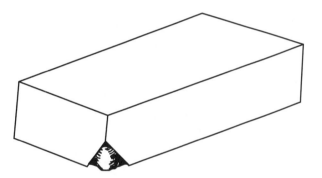

Example:

The Peek Box above has 10 marbles in it that are either red or blue. Only 1 marble may be viewed at a time through the cut-off corner. Students have shaken the box 10 times and recorded a tally of 2 red and 8 blue marbles. They are now to guess how many marbles of each color are in the box, and to explain their reasoning. The teacher might then ask, "If your first answer is not the correct one, how many of each color marbles do you think could be in the box, and why?" When all students have had a turn discussing their predictions and their reasons, open the box to allow students to count the actual number of marbles of each color.

Peek Box Probability

239

Extensions:

1. When students are ready, try Peek Boxes with 15 or 20 marbles in them. Tell the learners how many marbles are in the box and inform them that they will only be allowed 10 shakes. Have them complete their 10 shakes and tallies, make their predictions, and then discuss their tentative conclusions. (*Note:* Bridge a real-life connection with this activity by telling students that many times, when populations to be surveyed are very large, only a limited portion of the entire number can be sampled. Predictions as to totals or outcomes are often based on such probability samples.)

2. For advanced students, try Peek Boxes that contain marbles of 3 or 4 different colors, and perhaps as many as 100 marbles. As in Extension 1, allow only a limited sampling.

Peek Box Records

Total Number of Marbles = _____	(Tallies)
Marble Color #1 = _____ . I got this color ____ times.	
Marble Color #2 = _____ . I got this color ____ times.	
So I think there are ____ _____ marbles ____ _____ and marbles in the box.	
But, there could also be ____ _____ marbles and ____ _____ marbles in the box.	

Peek Box Records

Total Number of Marbles = _____	(Tallies)
Marble Color #1 = _____ . I got this color ____ times.	
Marble Color #2 = _____ . I got this color ____ times.	
Marble Color #3 = _____ . I got this color ____ times.	
Marble Color #4 = _____ . I got this color ____ times.	
So I think there are ____ _____ marbles and ____ _____ marbles in the box.	
But there could also be ____ _____ marbles and ____ _____ marbles in the box, **or** there could even be ____ _____ marbles and ____ _____ marbles in the box.	

Chapter 64

A Problem-Solving Plan

Grades K–8

☒ Total group activity
☒ Cooperative activity
☒ Independent activity
☒ Concrete/manipulative activity
☒ Visual/pictorial activity
☒ Abstract procedure

Why Do It:

Students will learn a method for analyzing and solving both word and other types of problems.

You Will Need:

Pencils and photocopies of the "Problem-Solving Plan" workpage provided will be necessary.

How To Do It:

1. At the outset, use the Problem-Solving Plan to talk through several examples of problems with the students. Then, after they have been oriented to the plan, have students use the step-by-step procedure outlined in the plan each time they encounter difficulty with a word problem or another problem-solving situation.

2. In general, do not expect students to complete all of the steps of the Problem-Solving Plan for every problem they encounter. They should, however, use the plan as a means for getting started and working through

the "tough spots" in problems. The Problem-Solving Plan also can serve as a diagnostic tool to help isolate the step or steps of a problem on which an individual frequently has trouble.

3. Of particular importance is the step titled "Select a Strategy." At this step, each participant must decide whether it might be helpful to try one or more of the following strategies:

* Build a Physical Model	* Take a Sample
* Act It Out	* Do a Simpler but Similar Problem
* Draw a Picture or Diagram	* Work Backward
* Make a Graph	* Guess and Check
* Make a Table	* Use Logical Reasoning
* Find a Pattern	* Use a Formula
* Classify	* Write an Equation
* Make a List	* Verify Your Work

4. Finally, encourage students to share and talk about the different procedures and reasoning each may have used. Such sharing should include both their answers and their thinking.

Example:

Students might use the Problem-Solving Plan to answer the problem that follows.

Six people were playing darts. Each threw 4 darts, and every dart hit the target shown. After adding their own point totals, they reported their scores as follows:

Jose = 19	Lisa = 28	Julie = 21
Dan = 30	April = 12	Jerry = 37

Dan said, "Some of these scores are not possible." Was Dan right or wrong? Which scores are possible and which are not? Explain why this is true.

The following is a sample Problem-Solving Plan of the kind students are to use to determine and explain which scores are possible and which are not.

A Problem-Solving Plan

Main Idea (In Your Own Words)

Some people are playing darts.

Question(s)

Are some, all, or none of the scores possible? How can we "prove" this?

Important Facts

- Each person threw 4 darts.
- All of the darts hit the target.
- Scores on the target = 1, 3, 5, 7, or 9.
- The players claimed totals of 12, 19, 21, 28, 30, and 37.

Select a Strategy

We might:

- Guess and Check
- Act It Out
- Find a Pattern

Solve It

By Guessing and Checking we found that 12, 28, and 30 would work. Also, 37 is too large (since 4 darts × 9 points = 36 points). Then we noticed that all of the totals that worked were even numbers, so we decided to try to Find a Pattern. We found a pattern!

Answer Sentence

The scores 12, 28, and 30 are possible, but 19, 21, and 37 are not.

Explain Why

We found the following pattern:

$$
\begin{array}{ll}
\text{Odd numbers} & \text{(Target scores)} \\
\underline{\times\ \text{Even numbers}} & \text{(4 darts per player)} \\
\text{Even numbers} & \text{(Final scores)}
\end{array}
$$

Extensions:

Continuing with the Example above, students might be asked to explain what happens when the following changes are made. These changes to the target problem are samples of extensions to a specific problem, but the idea of extensions can be used in any problem students might solve using the Problem-Solving Plan.

1. The same target is used, but an odd number of darts are thrown.
2. The target scores are all changed to even amounts and even numbers of darts are thrown.
3. Some of the target scores are even and some are odd. Students can be asked to explain what final scores will result.

A Problem-Solving Plan

Main Idea (In Your Own Words)
Question(s)
Important Facts
Select a Strategy
Solve It
Answer Sentence
Explain Why

Chapter 65

Fraction Quilt Designs

Grades 1–8

☒ Total group activity
☒ Cooperative activity
☒ Independent activity
☒ Concrete/manipulative activity
☒ Visual/pictorial activity
☒ Abstract procedure

Why Do It:

Students will use logical-thinking skills in the investigation of real-life geometry problems.

You Will Need:

This activity requires construction paper of several different colors, 1-inch graph paper, rulers, pencils, scissors, and glue.

How To Do It:

In this activity, students will represent fractional areas while designing a patterned quilt.

1. Students must first find as many ways as possible to split a square into congruent halves. You can use any size square, but a 4- by 4-inch square works well. Use the figure below to see some designs that show a dark region and a light region with equal areas. The students should cut their proposed solutions from construction paper and match them to check for congruence, and

then glue them back together on a larger piece of paper for display purposes. Be sure students have the opportunity to compare, contrast, and discuss the varied designs.

(EACH OF THESE DESIGNS SPLITS A SQUARE INTO 1/2S.)

2. The students next will design a 2- by 2-inch quilt pattern, with 1/2 in one color and the other 1/2 in another. To do this the students will need a 2- by 2-inch piece of graph paper and two colors of construction paper. After planning and marking the colored paper, they will cut out and glue the pieces on the graph paper. When finished they should be asked to ''prove'' that each color covers exactly 1/2 of the quilt design. They should also analyze how much area of the graph paper individual pieces cover, such as 1/4, 1/8, 1/16, and so on. (See the Example below.)

3. Successive activities will call on the students to design 3- by 3- and 4- by 4-inch square quilt patterns. They might also complete rectangular quilt patterns, such as 3 by 6 and 4 by 8 inches. When working with any of these patterns, the students should explore such factors as the many fractional amounts and equivalencies represented therein, the areas of their patterns as measured in square units, the many quilt patterns that are possible, and the significance of quilts in society.

Example:

The 2- by 2-inch quilt design shown here portrays several fractions and can be used to show fraction equivalence. The large triangles with ducks in them each make up 1/8 of the entire quilt. Since there are 4 such triangles, they make up 4/8 or 1/2 of the total quilt. The striped triangles each account for 1/16 of the total quilt; thus, the 4 of them equal 4/16 or 1/4 of the total quilt area (or half as much as the duck triangles). The smallest triangles, some of

Investigations and Problem Solving

which are solid black and some dotted, are each 1/32 of the total quilt area. (*Note:* Students might be asked, for example, how many of the smallest triangles it would take to make up 1/2 the quilt area, or what fraction would result if the smallest triangles were cut in half.)

Extensions:

1. Create a class quilt design on a large piece of butcher paper, with each student responsible for making a 1-square-foot block. Use several colors and a variety of design patterns. Post the finished quilt design on a large bulletin board or wall, and spend time discussing the fractional amounts, areas occupied by the different segments, the significance of the design chosen, and so on.

2. Students can read books about quilt making or do some research about the significance of quilts in society. The following are potential sources:

   Hawley, M'Liss Rae. *Scrappy Quilts.* Lafayette, Calif.: C & T Publishing, 2008.

   Horton, Roberta, and others. *Kaffe Fassett's: Quilts in the Sun.* Newtown, Conn.: Taunton Press, 2007.

   Johnson, Tammy, and Avis Shirer. *Folk-Art Favorites.* Woodinville, Wash.: Martingale & Company, 2009.

   Liddell, Jill. *The Changing Seasons: Quilt Patterns from Japan.* New York: Dutton Books, 1992.

   Linsley, Leslie. *The Weekend Quilt.* New York: St. Martin's Press, 1992.

   McNeill, Suzanne. *Batiks, Inspired by Bali.* Fort Worth, Tex.: Design Originals, 2009.

3. Interesting fraction questions might be asked about the 3- by 3-inch quilt block here, such as "What fraction of the total block is the central square?" and "When all of the white portions are combined, what fraction of the total quilt area do they make up?"

Fraction Quilt Designs

Chapter 66

What I Do in a Day

Grades 2–6

☒ Total group activity
☐ Cooperative activity
☒ Independent activity
☐ Concrete/manipulative activity
☒ Visual/pictorial activity
☒ Abstract procedure

Why Do It:

This activity helps students understand the properties of a circle graph using a real-life application.

You Will Need:

Each student should have a strip of paper divided into 25 sections with the end section shaded (the sample strip provided can be enlarged on a copy machine); a copy of the "Circle Graph Activity Sheet"; markers or crayons; tape or glue; a ruler; and a pencil.

How To Do It:

1. Begin by asking everyone to think about what they do on an average day. Give an example of how many hours a person might spend sleeping, eating, going to school, or doing homework. Then have students decide what activity will go in their "Keys." The Key is a table up in the right-hand corner of the activity sheet that states the activity, how many hours they spend on that activity, and what color that activity will represent on their graph. Be sure that the hours they show in their Keys add up to 24 hours in a day. Also, the Key can be extended to make room for more activities.

2. Learners will now color in their strips of paper according to their Keys. For example, if a student spends 8 hours sleeping, and the color for sleeping is red, he or she should color 8 squares on their strip red. If he or she spends 2 hours doing homework, and the color for homework is yellow, 2 squares on the strip must be colored yellow. This continues until the student has colored in the entire strip.

3. After learners have completely colored their strips, they will be ready to make their circle graphs. First, each strip is taped or glued together by making a loop with the strip and gluing or taping the shaded end to the back of the other end. A colored loop should be formed.

4. Now each student places his or her loop on the circle drawing on the "Circle Graph Activity Sheet" such that the center of the loop is the center of the circle and the colors show on the outside. A pair of concentric circles (circles with the same center) is formed.

5. The student then lines up one end of a color on the loop with the radius showing on the circle graph. He or she proceeds to go around the loop and makes a mark with a pencil on the circle at the end of each color, being sure to indicate what color will go in each section on the circle graph. See figure below.

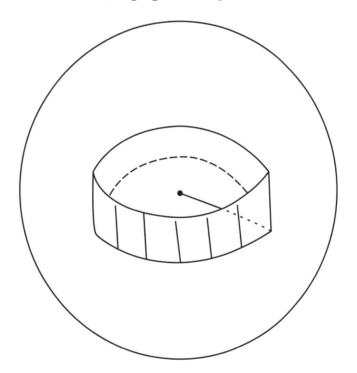

6. Finally, the student removes the loop from the graph and uses a ruler and pencil to connect the center of the circle with each mark made for each section on the circle graph. The student then is to color each section appropriately.

What I Do in a Day

Extensions:

1. A display of all the circle graphs displayed in the classroom can lead to an interesting discussion about everyone's day.

2. Students can discuss other ideas for circle graphs and decide on another one to do together, or each student could decide to do one on whatever he or she chooses.

3. Students could also make other graphs using their 24-hour day information, such as a bar graph or line graph.

Circle Graph Activity Sheet

24-Hour Strip

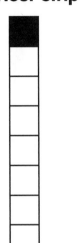

Color	Hours	Activity

What I Do During One Day (24 Hours)

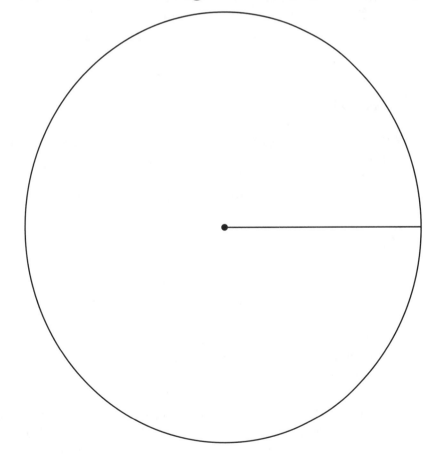

Chapter 67

Shaping Up

Grades 2–8

☒ Total group activity

☒ Cooperative activity

☒ Independent activity

☒ Concrete/manipulative activity

☒ Visual/pictorial activity

☒ Abstract procedure

Why Do It:

Students will learn to recognize the characteristics associated with an object, understand the idea of a set, and enhance logical-thinking skills needed for studying geometry and algebra.

You Will Need:

This activity requires photocopies of the "Attribute Pieces" page provided in two or three different colors (copying on card stock works the best) for each student or group of students, along with scissors, butcher or poster paper, and pencils.

How To Do It:

In this activity, students will use attribute pieces, which are objects that have more than one characteristic (for example, buttons can have such attributes as color, shape, number of holes), to learn about the study of sets. Sets are used throughout mathematics for organizing, classifying, and solving problems.

1. Students should cut out their attribute pieces and start by organizing them into piles any way they want, but they should be able to explain a reason for this

organization. For younger students use only two colors, and for older students use three colors. Or start the entire group with two colors and challenge them later to use three colors. Ask the students to describe their piles in words. For example, a student might say, "I have all the red pieces in this pile" or "I have all the small pieces in this pile." After students have organized their pieces in various ways, discuss the three different attributes these pieces have: color, shape, and size.

2. Next, with everyone listening, ask a student to describe one piece in words. For example, he or she might say, "This is a small, yellow circle."

3. Now ask the students to find the "Mystery Block" based on a series of clues, having them hold up the corresponding attribute piece that fits all the clues. Some possible clues are given in Example 1.

4. After a few Mystery Block problems, the learners are ready to describe a set of attribute pieces. Start by giving them the following sets: C = {all circles}, R = {all red pieces}, and L = {all large pieces}. Next, guide the students through a problem, for example by saying, "Use words to describe the set C − R." Then instruct learners to collect all the circles and all the red pieces. To find the set C − R, they are to start by collecting all the circles in one group and discarding anything that is not a circle. Then they take away all the red circles from the group remaining. Finally, students fill in the statement with the correct words, C − R = {all _____ and _____ circles}. The correct answer will be C − R = {all yellow and blue circles}. More problems are provided in Example 2.

5. To continue with the study of sets and attributes, give each group of students a large piece of paper to draw two overlapping circles as shown below. This is called a Venn Diagram.

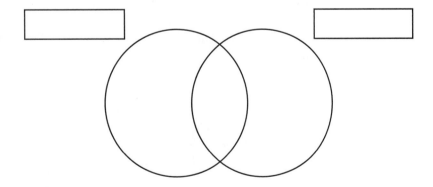

Using pencils, students will label one circle "Red" and the other circle "Triangle." Then instruct students to put all their attribute pieces into the Venn Diagram, leaving out any pieces that do not fit in either circle. It may be necessary to guide students through this problem to help them understand. For example, first tell them to find all the red pieces and put them in the circle labeled Red. Then have them find all the triangles and put them in the circle labeled Triangle. Then ask the learners what might go in the middle where the circles overlap. This should lead students to answer that the pieces in the overlapped section are each both red and a triangle. More possible labels for the circles are "Red and Circles," "Large and Squares," "Red Triangles and Large," "Not Yellow and Circles," "Not Blue and Small," or "Not Circles and Yellow."

6. Lastly, the students will put together a "difference train." This is a difficult concept for students to grasp and should be done in groups, so that students can discuss the problem and solution. Ask learners to put together a train of five pieces. Each piece should be different from the piece before it by only one attribute, as shown below.

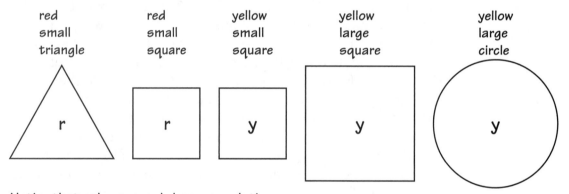

red	red	yellow	yellow	yellow
small	small	small	large	large
triangle	square	square	square	circle

Notice that only one word changes each time.

Examples:

Have students attempt the following sample problems.

1. Use only two colors—yellow and red. There is one piece that fits the clues.

1.	2.
Clue 1: It has 3 sides.	Clue 1: It is small.
Clue 2: It is yellow.	Clue 2: It is not yellow.
Clue 3: It is big.	Clue 3: It is round.

Answers: 1. This piece is a big, yellow triangle. 2. This piece is a small, red circle.

Use all three colors—yellow, red, and blue. There are two pieces that fit the clues.

1.	2.
Clue 1: It is not a square.	Clue 1: It is not red.
Clue 2: It has straight sides.	Clue 2: It has four sides.
Clue 3: It is not blue or red.	Clue 3: It is small.

Answers: 1. Small and large yellow triangle. 2. Small, yellow square and small, blue square.

2. Describe in words the following sets.

   1. L − R = {all large _____ and _____ pieces}

   2. R − C = {all red _____ and _____ pieces}

   3. R − L = {all _____, _____}

   4. C − L = {all _____, _____}

   5. L − C = {all large _____ and _____ pieces}

   Answers: 1. All large blue and yellow pieces. 2. All red squares and triangles. 3. All small, red pieces. 4. All small circles. 5. All large squares and triangles.

3. Use your attribute pieces and fill in the Venn Diagram shown below.

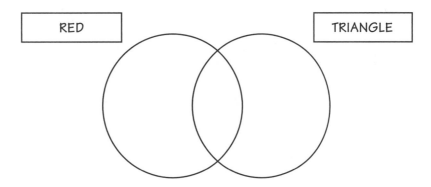

   Answer: All small and large red triangles go in the middle intersection. In the left section there should be all small and large red circles and squares. In the right section there should be all small and large, yellow and blue triangles.

Extensions:

1. A Venn Diagram with three circles overlapping can be used as shown below.

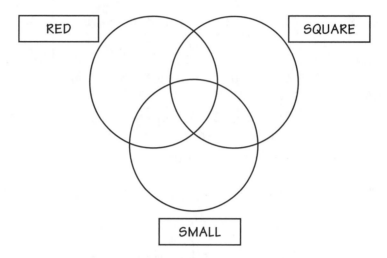

2. Students can be asked to put together a train of five pieces with a difference of two attributes.

3. *Shaping Up* can be done with alternative geometric shapes or different objects (such as buttons that have three or more different attributes).

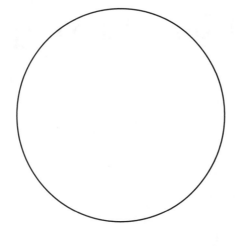
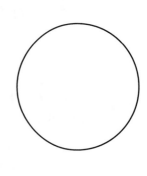
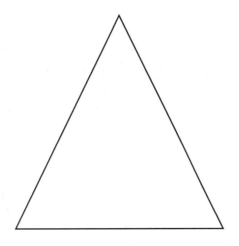
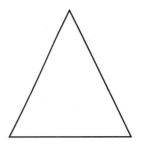

Chapter 68

Verbal Problems

Grades 1–8

☒ Total group activity

☒ Cooperative activity

☒ Independent activity

☐ Concrete/manipulative activity

☐ Visual/pictorial activity

☒ Abstract procedure

Why Do It:

Students will learn how to quickly analyze important problem information, exercise mental math skills, and work with problem-solving situations that occur outside the classroom.

You Will Need:

A selection of verbal problems are necessary; many possibilities for these are included below for young students (grades 1–3), middle grade students (grades 4–5), and older students (grades 6–8).

Directions and Problems for Young Students (Grades 1–3)

Directions:

- This is an exercise in listening as well as in working with numbers.
- I will read to you five questions.
- No grades will be taken on these questions. You will check your own answers.
- Number your paper from 1 to 5.
- Listen to the question carefully, think of the answer, and write only the answer on your paper.

Problems:

1. Karen has 2 dolls. Cheryl has 1 more doll than Karen has. How many dolls does Cheryl have? (*3 dolls*)

2. David has 4 toy cars. Luis has 3 toy cars. How many toy cars do both boys have? (*7 cars*)

3. John has 5 pieces of gum. Steven has 6 pieces of gum. Which boy has more pieces of gum? (*Steven*)

4. Nancy is 43 inches tall. Maria is 40 inches tall. Which one is taller? (*Nancy*)

5. Larry went to the store and bought 5 apples. On the way home, Jim gave Larry 1 apple. How many apples did Larry have when he got home? (*6 apples*)

6. Mary has 5 crayons in her box. Later the teacher gave her a yellow, an orange, and a purple crayon. How many crayons does she have now? (*8 crayons*)

7. John was asked to sharpen 10 pencils. Cheng was asked to sharpen 6 pencils. Which boy has to sharpen more pencils? (*John*)

8. Ann has 3 cookies. Her mother gave her 2 more cookies. How many cookies does Ann have? (*5 cookies*)

9. Mark has a stick that is 7 inches long. Jim has a stick that is 9 inches long. Which boy has the longer stick? (*Jim*)

10. Sally brought 4 dolls to the tea party, and Jane brought 3 dolls. How many dolls did they have at the party? (*7 dolls*)

11. Tom had 25 marbles and he gave 10 to his brother. How many did Tom have left? (*15 marbles*)

12. Mrs. Garcia needs 100 napkins. If she already has 70, how many more does she need? (*30 napkins*)

13. Mary has 2 birds and 11 fish. How many pets does she have? (*13 pets*)

14. There are 20 students in our class. If 1/2 of them are absent, how many are present? (*10 students*)

15. Spark can bark 10 times without stopping. Larky can bark 8 times without stopping. How many more times can Spark bark than Larky can bark without stopping? (*2 more times*)

16. If Ann brings 20 cookies and Kathy brings 10 cookies, how many cookies will they be bringing together? (*30 cookies*)

17. Joe has 2 pieces of cake and Mai has 4 pieces of cake. How many pieces do they have altogether? (*6 pieces of cake*)

18. Jackie had 11 marbles and gave 3 to her little brother. How many marbles does Jackie have left? (*8 marbles*)

19. Linda has 5 dolls and Marta has 6 dolls. How many dolls do they have altogether? (*11 dolls*)

20. Ken had 2 marbles. He won 5 more and then lost 3. How

many marbles did he end up with? (*4 marbles*)

21. Sally's mother baked 12 cupcakes. Sally and her friends ate 7 of them. How many cupcakes are left? (*5 cupcakes*)

22. Pablo wants to buy a pencil that costs 15¢. He has 8¢. How much more money does Pablo need? (*7¢*)

23. Karen has 3 pieces of candy, Sue has 2 pieces of candy, and John has 5 pieces of candy. How many pieces of candy do they have altogether? (*10 pieces of candy*)

24. Jorge has 2 dimes, 3 nickels, and 1 penny in his pocket. How much money does he have? (*36¢*)

25. Sam has 2 dogs, 3 goldfish, and 1 cat. How many animals does he have? (*6 animals*)

26. Mike threw the ball 9 feet and Ken threw the ball 14 feet. How much farther did Ken throw the ball than Mike? (*5 feet*)

27. Bill made 12 model airplanes. He gave 3 to John. How many did Bill have left? (*9 model airplanes*)

28. Sue put 12 balloons into groups of 4 each. How many groups of balloons did Sue have? (*3 groups*)

29. Diego bought one hot dog which cost 75¢. He paid the man with a one-dollar bill. How much change did Diego get back? (*25¢*)

30. Tom had 12 red cars and 8 blue cars. How many more red cars than blue cars did Tom have? (*4 more red cars*)

31. It is 8:00 A.M. and Tom must be at school in 25 minutes. At what time will Tom have to be at school? (*8:25 A.M.*)

32. Mr. Moreno had 5 bowls, 4 plates, and 4 saucers. How many dishes did he have in all? (*13 dishes*)

33. Mr. Brown has 4 rows of tulips with 3 tulips in each row. How many tulips does he have in all? (*12 tulips*)

34. Mrs. Chang paid $8 for 4 greeting cards. How much did each card cost? (*$2 each*)

35. If Johnny has a bag with 10 gum drops, and if he stops at the store and buys 6 more and then eats 2 on the way home, how many gum drops will Johnny have left? (*14 gum drops*)

36. Mrs. Davis is having 12 guests for dinner. If she has a loaf of bread with 24 slices, how many slices can Mrs. Davis serve each guest? (*2 slices*)

37. Farmer Brown has 4 chickens, and each chicken lays 2 eggs each day. How many eggs does Farmer Brown collect in one day? (*8 eggs*)

38. The elevator man went up 7 floors and down 3. What floor was he on if he started on the 1st floor? (*5th floor*)

39. Bill weighs 85 pounds. When he goes to camp for the summer, he loses 7 pounds

at camp. How much does Bill weigh when he goes back to school? (*78 pounds*)

40. If Elena has 5 dolls and she loses 2 dolls but later finds 1 doll, how many dolls are still missing? (*1 doll*)

41. A mother hen has 4 black chicks and 5 yellow chicks. How many chicks does she have in all? (*9 chicks*)

42. There are 3 goldfish in our aquarium. How many more do we need to buy so we will have 10 fish? (*7 fish*)

43. The mother bird raised two families this spring. In one family there were 3 babies. In the second family there were only 2. How many babies did the mother bird raise? (*5 baby birds*)

44. We are going to have company for dinner tonight. There will be 5 guests and our family of 6. How many plates will we need? (*11 plates*)

45. Marcia and John are gathering eggs. They have 7 eggs in their basket. How many more will they need to find to have a dozen eggs? (*5 eggs*)

46. If Tomas was to take 10 books and put them into 2 even piles, how many books would be in each pile? (*5 books*)

47. Roger weighs 7-1/2 pounds while Bill weighs 2-1/2 pounds less. How much does Bill weigh? (*5 pounds*)

48. In one of our reading groups we have 10 children. We have only 7 workbooks. How many more workbooks do we need so everyone has one? (*3 workbooks*)

49. Tom worked 12 arithmetic problems. If 8 of them were hard, how many were easy? (*4 problems*)

50. Mother hen has 7 chicks, and 5 of these chicks are black. The others are yellow. How many chicks are yellow? (*2 chicks*)

Directions and Problems for Middle-Grade Learners (Grades 4–5)

Directions:

- This is an exercise in listening as well as in arithmetic problem-solving skills.

- I will read to you ten questions. Odd-numbered questions, such as 1, 3, 5, and so on, are easier than the even-numbered questions. You may do only the odd- or even-numbered questions, or both if you wish.

- No grades will be taken on these questions. You will check your own answers.

- Number your paper from 1 to 10. Remember that you may choose to do only odd (easier) or even (harder) questions, or both. Challenge yourself!

- Listen to the question, think of the answer, and write only the answer on your paper.
- The questions will be read only once. Listen carefully.

Problems:

1. Mother made one dozen cookies. If Paul ate 9, how many would be left? (*3 cookies*)

2. Three boys went to the store to buy bubble gum. Oscar bought 8 pieces, Willy bought 15 pieces, and Jonas bought 12 pieces. How many pieces did they buy altogether? (*35 pieces of gum*)

3. Roberto had 23 marbles. He won 9 more in a game. How many did he have altogether? (*32 marbles*)

4. There are 33 students in one third-grade class, and 29 in another. How many students are there in both classes? (*62 students*)

5. Sally had 15 apples. She ate 2 and gave 6 away. How many did she have left? (*7 apples*)

6. There are 29 children in Mrs. Suzuki's third-grade class. If 16 are boys, how many are girls? (*13 girls*)

7. Bill and Jim went to the rodeo on Saturday. They saw 8 white horses and 3 black horses. How many more white horses did they see than black horses? (*5 more white horses*)

8. Jane brought 2 pints of lemonade to the Thanksgiving party, and Susan brought 1 pint of lemonade. Each pint contains 2 cups. How many cups of lemonade could they serve at the party? (*6 cups*)

9. Maria went to the grocery store for her mother. She bought 3 boxes of cookies. There were 8 cookies to the box. How many cookies did she buy? (*24 cookies*)

10. At the end of the sixth inning, the score at the baseball game was 8 for the Red Sox and 5 for the Tigers. In the last inning the Red Sox made 4 runs, and the Tigers made 6 runs. Which team won the game? By how many runs? (*Red Sox by 1 run*)

11. John went to Mr. Lang's orchard to pick apples. If one bushel of apples weighed 50 pounds, how many would 4 bushels weigh? (*200 pounds*)

12. Mary has 3 skirts and 4 blouses. How many outfits can she make by using different blouses with each skirt? (*12 outfits*)

13. A pint is 1/8 of a gallon. How many gallons is 10 pints? (*1-1/4 gallons*) 24 pints? (*3 gallons*) 33 pints? (*4-1/8 gallons*)

14. Mrs. Rivera went shopping and bought $12.48 worth of groceries. If she bought 12 items, what was the average cost of each item? (*$1.04 each*)

15. The distance from Stockton to Lodi is 22-1/2 miles. How many miles is the round trip? (*45 miles*)

16. If you saved 7¢ of every 20¢ that you earned, how much money would you have saved after you had earned 60¢? (*21¢*)

17. Ted and John bought a Christmas tree for their parents. Ted wanted to buy a tree that was 3 feet 7 inches tall. John wanted to buy a tree that was 4 feet 6 inches tall. They decided to buy the tree John had picked out. How many inches taller than Ted's tree is John's tree? (*11 inches*)

18. Lecosha wanted to buy some ribbon for her new dress. She liked a yellow ribbon that was 21 inches long. She also liked a green ribbon that was 2 feet long. If she bought the longer one, which one did she buy? (*green ribbon*)

19. There are 12 apples on the table. Three girls want to share the apples equally. How many apples will each girl eat? (*4 apples*)

20. Three boys went fishing and they caught 21 fish. Bob caught 7 fish. Jerry caught 8 fish. How many fish did Kim catch? (*6 fish*)

21. Claudia bought 2 yards of material. How many inches of material did Claudia buy? (*72 inches*)

22. The bus left Stockton at 8:25 A.M. It arrives in Sacramento 1 hour and 25 minutes later. What time will it arrive in Sacramento? (*9:50 A.M.*)

23. Mary has 4 pies that she wants to cut into pieces so 12 people can have equal shares. How much pie will each person get? (*1/3 of a pie*)

24. When Chue took a trip, it took him 1/2 hour one way and 2/3 hour on the way back. How many minutes did his trip take? (*70 minutes*)

25. John has 7 cookies and Stan has 8. They wanted to divide them into 5 groups for their friends. How many cookies did each friend get? (*3 cookies*)

26. A rug is 4 feet wide and 12 feet long. What is its area? (*48 square feet*)

27. Harry walked 3-3/4 miles in the morning and 2-1/4 miles in the afternoon. How far did he walk altogether? (*6 miles*)

28. Omar has 59¢. How many tickets at 5¢ each can he buy? (*11 tickets with 4¢ left*)

29. Six classrooms are to share equally in a shipment of 42 new kickballs received at Terry School. How many kickballs will each classroom receive? (*7 kickballs*)

30. Karen has 54 photographs taken at Bass Lake last summer. She can put 6 photos on a page in her photo album. How many pages

will she fill with 54 photographs? (*9 pages*)

31. Jim practiced on his trumpet for 25 minutes on Tuesday and 15 minutes on Wednesday. How many total minutes did he practice? (*40 minutes*)

32. Janice spent 35¢ for a soda each day. How much did it cost her for 5 days? (*$1.75*)

33. Betty must ride a bus to school. She walks 3/4 mile to the bus stop. When she gets on the bus, she rides another 2-1/4 miles to school. How far does Betty live from school? (*3 miles*)

34. Dennis has 44 boxes all alike in a wagon. The total weight of all the boxes is 132 pounds. How much does each box weigh? (*3 pounds*)

35. Each person in Ms. Wilson's class will get 5 pieces of paper. If there are 30 children in the class, how many pieces of paper will Ms. Wilson need? (*150 pieces of paper*)

36. If a small box of apples costs $2.50, how much will 4 boxes cost? (*$10*)

37. Mr. Gomez had 800 peaches to pack in boxes. If he puts 20 peaches in each box, how many boxes will he need? (*40 boxes*)

38. Sandra had 22 pieces of candy and received 5 more. She then gave 17 pieces away. How many pieces of candy did Sandra have left? (*10 pieces of candy*)

39. How much change will Mary receive from her 75¢ after she buys a pencil for 15¢, paper for 16¢, and candies for 35¢? (*9¢*)

40. At a Halloween party, 35 children were grouped in 3s to play a game. How many complete groups of 3 were there? (*11 complete groups*)

41. Mary has 28 paper dolls. How many will she give away if she gives her sister half of them? (*14 paper dolls*)

42. Manuel placed 16 chairs in each row in the music room. How many chairs did he place in 3 rows? (*48 chairs*)

43. Ann bought a banana for 39¢. She gave the clerk 50¢. How much change did she receive? (*11¢*)

44. The Tanakas are traveling 300 miles from the lake to their home. They have gone 248 miles of this journey. How many miles have they still to go? (*52 miles*)

45. In the number 8,621, what is the value of 2? (*2 tens or 20*)

46. Two quarts equal how many pints? (*4 pints*)

47. Which is smaller, 1/8 or 1/16? (*1/16*)

48. A pie is cut into 8 equal parts, and John eats two of them. What fractional part of the pie is left? (*3/4 or 75%*)

49. A baseball team needs 9 players. How many baseball teams can be made up from 27 players? (*3 teams*)

50. Jorge has 88 pennies, which he wants to exchange for nickels. How many nickels can he get for them? (*17 nickels plus 3 pennies or 17.6 nickels*)

Directions and Problems for Older Learners (Grades 6–8)

Directions:

- This is an exercise in listening as well as in arithmetic problem-solving skills.

- I will read to you ten questions. Odd-numbered questions, such as 1, 3, 5, and so on, are easier than the even-numbered questions. You may do only the odd- or even-numbered questions. You may do both if you wish.

- No grades will be taken on these questions. You will check your own answers.

- Number your paper from 1 to 10. Remember you may choose to do only odd (easier) or even (harder) questions, or both. Challenge yourself!

- Listen to the question, think of the answer, and write only the answer on your paper.

- The questions will be read only once. Listen carefully.

Problems:

1. Jack paid 90¢ for 3 special stamps. How much did each stamp cost? (*30¢*)

2. Mr. Perez's horse is 15 hands high. A hand is 4 inches. How many feet high is the horse? (*5 feet*)

3. Tom had 25 marbles, Tim had 50 marbles, and Joe had 100 marbles. How many more marbles did Joe have than Tom? (*75 marbles*)

4. A special express train in Japan travels 320 miles between Tokyo and Osaka at 160 miles an hour. How many hours does the trip take? (*2 hours*)

5. If 36 children are grouped into teams of 9 each, how many teams will there be? (*4 teams*)

6. A company of soldiers marched 40 miles in five days. The first day they marched 9 miles; the second day, 10 miles; the third day, 6 miles; the fourth day, 8 miles. How many miles did they march on the fifth day? (*7 miles*)

7. Jose had 24 papers to sell. He sold 9 of them. How many papers has he left to sell? (*15 papers*)

8. One gallon of gasoline weighs 5.876 pounds. What will 10 gallons of gasoline weigh? (*58.76 pounds*)

9. Texas has an area of approximately 260,000 square miles, and California has an area of approximately

160,000 square miles. How much larger is Texas than California? (*100,000 square miles*)

10. Jan bought a cap for $5.25 and a scarf for $1.50. She gave the clerk a ten-dollar bill. How much change did she receive? (*$3.25*)

11. Albert saved $15.98. He spent all but $1.98 of it on Christmas gifts. How much did he spend on Christmas gifts? (*$14*)

12. A restaurant owner paid $12.50 for a turkey priced at 50¢ per pound. What was the weight of the turkey? (*25 pounds*)

13. A small town has 100 parking meters. The average weekly collection from each meter is $4. What would be the total weekly collection? (*$400*)

14. At the equator, the Earth's surface moves about 1,000 miles per hour as the Earth revolves on its axis. If you lived at the equator, how far would you be carried in a complete day? (*24,000 miles*)

15. Oliver's father earned $120 for a 4-hour job. What was his hourly rate of pay? (*$30.00*)

16. The 5,000-mile trip from Seattle to Tokyo required 20 hours of flying time. What was the average speed in miles per hour? (*250 mph*)

17. Mario's father drives a bus. He has made 150 trips of 100 miles each. How many miles has he driven? (*15,000 miles*)

18. At 35 miles per hour, how long will it take to drive an automobile a distance of 210 miles? (*6 hours*)

19. The manager of a school store sold 100 dozen pencils. How many pencils did she sell? (*1,200 pencils*)

20. A traffic court showed that 615 cars passed a certain point in an hour. At this rate, how many cars would pass in 6 hours? (*3,690 cars*)

21. About $36,000 is spent each year for paint used on the Truckee Bridge. What is the average cost per month? (*$3,000*)

22. John delivers an average of 200 newspapers a week. At this rate, how many newspapers will he deliver in a year? (*10,400 newspapers*)

23. A pound of sugar will fill 2-1/4 cups. How many cups can be filled from a 2-pound package? (*4-1/2 cups*)

24. A paper company owns 4,000 acres of timberland. In order to increase its landholdings by 400%, how many additional acres must the company buy? (*12,000 acres*)

25. John and David want to share the cost of a model car kit that costs $3.00. How much will each boy have to pay? (*$1.50*)

26. A pilot estimating the gasoline needed for a flight allowed a margin of 25% of the total gas needs for the sake of safety. If the trip required 200 gallons of gas,

Investigations and Problem Solving

how many gallons were put into the tanks? (*250 gallons*)

27. Kathy wants to go horseback riding, which costs $1.50 for one hour. She can earn 50¢ an hour by babysitting. In how many hours of babysitting can she earn enough for one hour of riding? (*3 hours*)

28. The enrollment of a small college dropped 5% from a high of 1,000 students. What was the enrollment then? (*950 students*)

29. Ms. Garfolo and Ms. Bartell had 64 pupils between them. Ms. Garfolo had 40 pupils and Ms. Bartell had 24. In order for the teachers each to have the same number of pupils in her room, how many should each have? (*32 pupils*)

30. Lannie's father can get a $400 outboard motor with a reduction of $40. What percent is the reduction of the regular price? (*10%*)

31. Jerry's team scored the following scores in kickball this week: Monday, 3 runs; Tuesday, 4 runs; Wednesday, 0 runs; Thursday, 2 runs; Friday, 1 run. How many runs did Jerry's team score altogether? (*10 runs*)

32. Rene's baby brother must be given his bottle every 4 hours. If the baby was last fed at 11:30 A.M., what time will the baby need his next bottle? (*3:30 P.M.*)

33. David's dog eats a dog treat a day, and the treat costs 20¢. How much does it cost to treat the dog per week? (*$1.40*)

34. Ranger VIII took about 4,000 pictures of the moon during the last 10 minutes of flight. How many pictures a minute did the Ranger camera take? (*400 pictures*)

35. In arithmetic this week, Candy missed the following number of problems: 3, 4, 5, 1, and 2. How many problems did she miss this week? (*15 problems*)

36. The astronaut John Glenn orbited the Earth every 1-1/2 hours. How many orbits did he make in 4-1/2 hours? (*3 orbits*)

37. It takes Jim 5 minutes to walk to school. He also goes home for lunch each day. How much time does Jim spend each day in walking back and forth between school and home? (*20 minutes*)

38. Fire records showed that about 60 out of the last 150 fires were caused by sparks from other fires. What fraction of the fires were caused by such sparks? (*2/5*)

39. Mary Ann's mother told her to be home at 4:00 P.M. Mary Ann didn't get home until 5:10 P.M. How late was she? (*1 hour, 10 minutes*)

40. Juan's class picture costs $15.00 for the large picture. The individual pictures cost $1.00 each if he buys 12 of them. For how much should Juan's mother make the

Verbal Problems   **269**

check if he keeps them all? (*$27.00*)

41. Kathy's mother told her to bake a double recipe of cookies. This means that Kathy must double all of the measurements. The recipe calls for 1 cup of milk. Will Kathy need a pint or a quart of milk for her cookies? (*1 pint*)

42. At a market, a sign for apples read: 4 pounds for 20¢. If Mary bought 5 pounds of apples, how much would she have to pay? (*25¢*)

43. Jane bought a *Time* magazine for $4.00 and a *Seventeen* magazine for $3.50. How much did Jane have left out of her $10 allowance? (*$2.50*)

44. If Susan was 9 years old in 1984, how old was she in 1990? (*15 years old*)

45. How many hours of instruction are in a school day that begins at 9:00 A.M. and ends at 3:30 P.M., with an hour out for lunch? (*5-1/2 hours*)

46. Sam and Ethan were playing marbles. Sam began with 10 marbles and Ethan began with 12. At the end of the game Ethan had lost 3 of his marbles to Sam. How many marbles did Sam have? (*13 marbles*)

47. John wants a driving permit when he is 15-1/2 years. He is now 11-1/2 years. How long must he wait before he applies? (*4 years*)

48. If in 3 nights Sasha slept 10, 6, and 8 hours, respectively, what was the average amount of sleep she got per night? (*8 hours*)

49. In basketball Cincinnati had 30 wins and 12 losses. How many more wins than losses did Cincinnati have? (*18 wins*)

50. Four boys together bought 2-dozen cookies. They saved half of the cookies, and divided the rest evenly among themselves. How many cookies did each boy get? (*3 cookies*)

Chapter 69

Scheduling

Grades 3–8

☒ Total group activity
☒ Cooperative activity
☒ Independent activity
☐ Concrete/manipulative activity
☒ Visual/pictorial activity
☒ Abstract procedure

Why Do It:

Students will work through a real-life problem situation and develop organizational skills.

You Will Need:

Each student will need at least one copy of the "My Weekly Schedule" worksheet (provided) for planning purposes.

How To Do It:

1. Provide each learner with a copy of "My Weekly Schedule." Have students first fill in the chart spaces for the upcoming week with those activities that have designated times. Then, for any unfilled time slots, have them pencil in desired activities. Allow them to share and discuss their schedules with each other. You might also have them analyze their schedules in terms of the "wise use of time." (For example, if there is going to be a math test on Friday, it might not be wise to spend all of Thursday evening's unscheduled time watching TV.)

2. Students of different ages will have varying activities with which to fill their charts. Young students might,

for instance, spend 30 minutes getting ready for school in the morning and leave for school at a set time. Middle-grade students might help with family chores and earn spending money doing weekend jobs. Older students may have the greatest need to use their time wisely because they often have part-time jobs or other activities to attend outside school.

Example:

The students shown below are comparing and commenting on their personal schedules and solving the problem of when to put in a particular activity.

Extensions:

1. Have the learners develop a schedule for a week when they will not be in school. They might want to compare and contrast it with a school-week schedule.

2. Consider sharing a weekly or monthly lesson-plan schedule with the students. When you do so, you can point out not only what will be studied but also why it is important that certain things be learned in sequence.

3. Allow the students to do some long-term planning. Planning a monthlong period can often be very revealing. You might also want them to see some examples of yearlong plans (or even 5- or 10-year projections).

My Weekly Schedule

	Sunday	Monday	Tuesday	Wednesday	Thursday	Friday	Saturday
6:00 A.M.							
6:30 A.M.							
7:00 A.M.							
7:30 A.M.							
8:00 A.M.							
8:30 A.M.							
9:00 A.M.							
9:30 A.M.							
10:00 A.M.							
10:30 A.M.							
11:00 A.M.							
11:30 A.M.							
12:00 P.M.							
12:30 P.M.							
1:00 P.M.							
1:30 P.M.							
2:00 P.M.							
2:30 P.M.							
3:00 P.M.							
3:30 P.M.							
4:00 P.M.							
4:30 P.M.							
5:00 P.M.							
5:30 P.M.							
6:00 P.M.							
6:30 P.M.							
7:00 P.M.							
7:30 P.M.							
8:00 P.M.							
8:30 P.M.							
9:00 P.M.							
9:30 P.M.							
10:00 P.M.							
10:30 P.M.							
11:00 P.M.							
11:30 P.M.							

Chapter 70

Student-Devised Word Problems

Grades 3–8
- ☒ Total group activity
- ☒ Cooperative activity
- ☒ Independent activity
- ☐ Concrete/manipulative activity
- ☐ Visual/pictorial activity
- ☒ Abstract procedure

Why Do It:

Students will create and use their own word problems based on everyday things that are of personal interest.

You Will Need:

Select or devise 3 or 4 word problems that every learner in the group will be able to solve quite easily. In addition, a copy of ''A Problem-Solving Plan'' and a list of the strategies for problem solving should be made available (see *A Problem-Solving Plan,* p. 242). The students will also need a supply of scratch paper; pencils; and some 5- by 8-inch index cards.

How To Do It:

Begin by asking the students to discuss and solve a simple word problem. The discussion and solution process should include using the Problem-Solving Plan and one or more of the strategies. After the students have solved the problem, guide the students through writing their own problem, by using the first problem as a guideline and leaving out certain words. Students then solve this new problem. Finally, the students write another problem using another similar guideline with more words left out. Again the students solve this new problem. Students continue this process of modifying and rewriting until they have, in fact, created an entirely new problem. The Example below shows the rewriting process for a given problem.

Example:

The word problem below has been modified several times until it has become a totally new problem, and one in which the students have a personal interest because they created it.

Original Word Problem:
Doug has 8 marbles and Susan has 11 marbles. Who has more marbles?

Note: The initial word problem should be easily solved by everyone in the group; obviously, advanced students would start with a more difficult problem.

1st Rewrite:

_____ has ____ marbles and _____ has ____ marbles. Who has more marbles?

Note: The names and amounts have been left blank so that the student may fill in his or her name and that of a friend, along with number amounts with which he or she feels comfortable working. Having done so, the learner should then solve the new problem.

2nd Rewrite:

_____ has ____ _____ and _____ has ____ _____ . Who has more _____ ?

Note: Now not only are the names and amounts to be changed but also the items being dealt with are to be replaced.

3rd Rewrite:

_____ has ____ _____ and _____ has ____ _____ . _____ ?

Note: Students are now required to change the question. (Such typical questions as How many more? or How many less? or What is the total? should be discussed.) Point out to the learners that they have now created their own, new word problem.

The learners will likely need to practice this word-problem rewriting process, but once they have mastered it they will begin to understand that all word problems have essentially the same components. At this point they will be ready to write word problems of their own with little or no assistance. To enhance this procedure in interesting ways, have the learners try some of the Extensions noted below.

Extensions:

1. Once students are familiar with how to write their own word problems, they might like to compose some more of their own. Remind them that, as they write them, their problems must contain the elements listed on the Problem-Solving Plan and they must be solvable. Then provide learners with scratch paper and let them begin. On one piece of scratch paper they should write their proposed word problems and on another calculate the answer. When an individual thinks she or he has completed a problem, she or he must take the problem to another student to have them try it. If they both agree the problem is OK and get the same answer, it may be a "good" problem; but if something seems askew, the two of them must sit together and edit the problem until it is workable. Then, in turn, they must have a third student try the problem. If the problem is OK it is shared with the teacher; if not, the three of them must edit it again.

   When finished writing and editing, the learner brings the problem to the teacher for a final check. If the problem is OK, the student is given a 5- by 8-inch index card and is directed to write the problem on the front of the card, in his or her best penmanship. The front of the card must also say "Authored by (Student's Name)," and it may be decorated (as with a picture frame or with horse drawings, if that is what the problem is about). Finally, "Solved by" must be written on the back of the card. The answers remain only at the author's desk. Once each student has written two or three such problems, allow a session during which they attempt to solve each other's word problems. When an individual thinks he or she has the solution, that student must go to the author's desk to check whether the answer is correct; if it is, he or she gets to write his or her name on the back of the card where it says "Solved by." Since these problems are personal, and about their friends, the learners will have a great time!

2. Have the students complete a word-problem writing process similar to that in Extension 1, but specify the numbers they must include. "Easy" numbers to use are those that fit together easily, perhaps 3, 4, and 12. "Tough" numbers might include 2, 17, and 512. Whatever numbers are suggested, students may incorporate these into the problem in a variety of formats, including, for example, (1) Seen Numbers—those that can be easily viewed when reading a problem; (2) Hidden Numbers—an answer might be considered to be a hidden number; (3) Numbers as Words—such as 3 written as three; (4) Important Information—numbers necessary for the solution of the problem; and (5) Extra Information—numbers not needed for solution.

Chapter 71

Tired Hands

Grades 2–8

☒ Total group activity
☒ Cooperative activity
☐ Independent activity
☒ Concrete/manipulative activity
☒ Visual/pictorial activity
☒ Abstract procedure

Why Do It:

Students will learn graphing and analyzing skills after collecting data from an experiment involving a physical activity.

You Will Need:

A watch or clock that displays seconds is required, as well as graph paper (or a group graph on the chalkboard or overhead projector) and pencils.

How To Do It:

1. Tell the students that they will be taking turns exercising their hands for 90 seconds (1-1/2 minutes) at a time. The exercise requires that one hand lie flat, palm up, with fingernails and wrist touching the table. The hand must then be clenched into a fist and opened, repeatedly. Allow students to practice this a few times, ensuring that they keep the backs of their hands on the table.

Time in Seconds	Count for Each 15 Seconds
0–15	
16–30	
31–45	
46–60	
61–75	
76–90	

2. Have students work in partners, with one person doing the exercising and the other recording the number of hand closings at each 15-second interval. The exerciser is to count out loud 1, 2, 3, 4, and so on, and the recorder must, when the teacher calls out ''Record,'' write down the total every 15 seconds (use a chart similar to the one shown above). It also helps if the exerciser, without stopping hand movement, begins the count anew for each segment. The activity continues in this manner for 90 seconds. The partners then change roles and the process is repeated.

3. When all have finished, each student should graph, analyze, and discuss the outcomes from his or her own hand exercise. The students might also like to plot everyone's records on a group graph, such as the one displayed in the Example, and compare results.

Example:

The following graph shows the results of the hand exercise for a group of students. This graph indicates some interesting trends. For example, nearly all students started at a fast rate, then about midway they seemed to become fatigued and slow down, but near the finish many made a last ''sprint'' to the end.

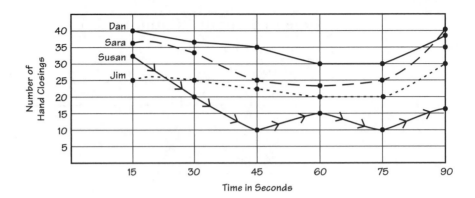

Extensions:

1. Suggest that the participants test their muscle recovery rates. To do so they should wait 30 seconds (after an initial 90-second test) and then exercise and keep records for another 30 seconds, then rest 30 seconds and exercise another 30 seconds, and so on. Have them compare their own rates and discuss what happened.

2. Have the learners complete the hand-exercise activity while using their nondominant hands. Then have them graph and compare the outcomes to those from their first experiences.

3. Have the participants design a test to find the tiring and recovery rates for their arm or leg muscles.

Chapter 72

Paper Airplane Mathematics

Grades 2–8

☒ Total group activity
☒ Cooperative activity
☒ Independent activity
☒ Concrete/manipulative activity
☒ Visual/pictorial activity
☒ Abstract procedure

Why Do It:

Students will apply problem solving, measurement, geometry, and logical thinking to a fun activity.

You Will Need:

This activity requires the "Airplane Contest Records" handout (provided), plain paper, other paper of assorted types and weights, rulers (metric or English), yardsticks or meter sticks, pencils, and scissors.

How To Do It:

1. Begin by having the students suggest the best ways to make paper airplanes. Then supply each of them with several sheets of typing paper. Tell them that they may construct as many airplanes as they wish, but only two may be entered in the upcoming Paper Airplane Contest, one in Category A and one in Category B. Category A is for those planes attempting to fly the farthest. Category B airplanes will try to land on a

target that is 10 yards (or meters) away. A further requirement (recommended for most students) is that they submit a plan before entering an airplane in the contest. The plan must include, among other important information:

- A sketch of the airplane
- Notation of special features
- Length of the plane
- Distances flown during testing

- Width at the tail
- Flight-test accuracy records
- Depth at the tail
- Area of the wing surfaces

2. Allow the participants to begin designing, folding, cutting, and modifying their paper airplanes. Any folded shape is allowed, and extra material may be cut off, but nothing can be added. Also provide time for flight testing. At a designated time, hold the Paper Airplane Contest and have each student keep records (see the "Airplane Contest Records" sheet). Following the contest, encourage discussion about which paper airplanes went the farthest (Category A), which were the most accurate (Category B), and why. The students may also be allowed to design "better" airplanes, which would compete in a follow-up contest. Students should summarize their findings from any event by answering the last questions on the "Airplane Contest Records" sheet.

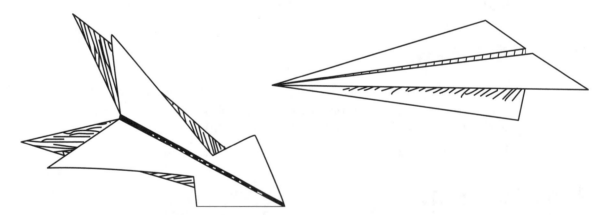

Example:

The features of the paper airplanes shown above are quite different from one another. Have students ponder such questions as Which airplane will fly farther? Which will be more accurate? and Which will stay aloft longer?

Extensions:

1. Consider adding a third category to the contest that gauges which paper airplane will stay aloft the longest. If included, a stopwatch will be a needed piece of equipment.

2. Students can test which airplane is capable of carrying the heaviest cargo over a specified distance, for example by using paper clips to weigh down the airplanes.

3. Allow the learners to experiment with airplanes built using paper of different types and dimensions.

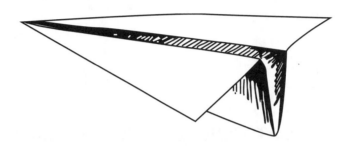

Airplane Contest Records

It is time for our Paper Airplane Contest. Get ready to tally the results. Keep a record of them on the following charts:

CATEGORY A: AIRPLANES FLYING THE FARTHEST

Name	Distance Flown	Features of the Plane

CATEGORY B: MOST ACCURATE AIRPLANES

Name	Distance from Target	Features of the Plane

Which of the paper airplanes flew the greatest distances? Did certain features seem to allow them to fly farther than other airplanes? If so, what features were they? Which airplanes were the most accurate in landing? Did they have features different from those flying long distances? If so, what were the differences?

_____

_____

_____

_____

_____

_____

_____

Investigations and Problem Solving

Chapter 73

A Dog Pen Problem

Grades 3–8

- ☒ Total group activity
- ☒ Cooperative activity
- ☒ Independent activity
- ☒ Concrete/manipulative activity
- ☒ Visual/pictorial activity
- ☒ Abstract procedure

Why Do It:

This activity will help students understand perimeter and area relationships by mapping and physically creating fenced areas.

You Will Need:

Several lengths of rope or string from 8 to 100 feet are required, as well as a 1-square-foot piece of cardboard for each participant; chalk; tape; pencils; graph paper; and such measuring devices as yardsticks, 1-foot rulers, and a measurement trundle wheel (optional).

WOW! THAT'S AN ODD SHAPE FOR A DOG PEN.

YOU WOULD THINK THE PEN SHOULD HAVE SPACE FOR THE DOG TO TURN AROUND IN! I THINK A SQUARE PEN WOULD GIVE THE DOG MORE ROOM.

How To Do It:

1. Help the participants review their knowledge of the terms *square, rectangle, square foot, perimeter,* and *area.* Then use a rope or string, of perhaps 12 feet, to demonstrate *A Dog Pen Problem.* Tape the rope to the floor or playground in a square and ask the students to determine the perimeter and the area. They should use yardsticks or foot rulers to measure the perimeter of this dog pen as 3 feet + 3 feet + 3 feet + 3 feet = 12 feet. The students should next place as many of their cardboard square feet inside the roped-off space as possible; they will find that the area of this pen = 9 square feet. Participants should use their graph paper to draw a map (see Example) of this dog pen, along with their measurements.

2. Next allow the students to use the same length of rope to create other rectangular dog pens. (*Note:* Initially allow only measurements in increments of 1 foot to make it easier.) They will soon discover that they can create a pen with a perimeter measuring 4 by 2 by 4 by 2 feet = 12 feet; have them use the cardboard squares to determine that the area = 8 square feet. They will also find, much to the surprise of some, that the same rope can be used to pen off a perimeter of 5 by 1 by 5 by 1 feet = 12 feet, but that this pen only has an area that = 5 square feet. These findings should also be mapped on graph paper.

3. Ask the participants to explain, in their own words, what effect the shape of a dog pen has on perimeter and area. Next, organize the students into working groups and have them construct, measure, and record their findings for a variety of dog pens. (Good rope lengths to work with are 8, 16, 20, 24, 36, 40, 60, and 100 feet.) When finished, have them share what they learned from the activity.

Example:

These dog pen maps all have perimeters of 12 feet, but their areas are quite different.

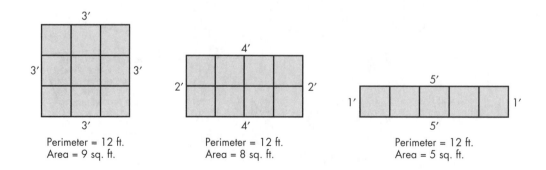

Perimeter = 12 ft.
Area = 9 sq. ft.

Perimeter = 12 ft.
Area = 8 sq. ft.

Perimeter = 12 ft.
Area = 5 sq. ft.

Investigations and Problem Solving

Extensions:

1. Challenge advanced students to try shapes other than rectangles. They might try designing dog pens in the form of triangles, circles, ovals, or other geometric shapes. (*Note:* When doing so, students may need to estimate the square footage.)

2. Ask the students to decide how their understandings of perimeter and area apply to other everyday activities. For instance, ask, "If carpet is usually sold by the square yard, what shape room will be the least expensive to carpet?" Or ask, "In what shape should a building be constructed to achieve the maximum inside volume while using the minimum surface area of building materials?"

Chapter 74

Building the Largest Container

Grades 3–8

☒ Total group activity
☒ Cooperative activity
☒ Independent activity
☒ Concrete/manipulative activity
☒ Visual/pictorial activity
☒ Abstract procedure

Why Do It:

This project provides learners with a hands-on, problem-solving experience that may be solved intuitively, logically, or both.

You Will Need:

A sheet of heavyweight paper or tagboard (file-folder stiffness or thicker) is needed for each student, as well as scissors, rulers, masking tape, a bag of rice (or other dry material that may be poured and measured), and a volume-measuring device (such as a large measuring cup or a scientific beaker).

How To Do It:

1. Provide each student with an 8-1/2- by 11-inch (or other standard size) sheet of tagboard. Explain that their job, using just one sheet of tagboard, is to construct the container that will hold the greatest volume of rice. They may measure, cut, bend, and tape the tagboard in any manner, but before doing so they should develop

a plan (either cooperatively or individually) that they think will yield the largest container.

2. Complete one or two sample containers (as in the Example below) for initial comparisons. This may get the students to begin to think about some questions, such as (1) Is a rectangular, box-like figure the best shape and, if so, should it have low or high sides? (2) What about a triangular shape? (3) Would a cylinder be better?

3. Allow the learners to plan and experiment. When each container is completed, its designer should fill it with rice and use the measuring device to determine its volume. Ask students to keep a record of the shapes and dimensions of the varying containers, and encourage them to compare and contrast their findings. Permit learners to make two or three containers if they wish to improve their designs.

Example:

The containers shown below both have been built from 8-1/2- by 11-inch tagboard, but the volume of rice that they will hold is quite different.

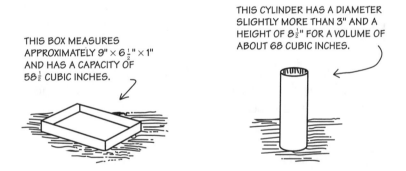

THIS BOX MEASURES APPROXIMATELY 9" × 6½" × 1" AND HAS A CAPACITY OF 58½ CUBIC INCHES.

THIS CYLINDER HAS A DIAMETER SLIGHTLY MORE THAN 3" AND A HEIGHT OF 8½" FOR A VOLUME OF ABOUT 68 CUBIC INCHES.

Extensions:

1. Younger students will likely need to complete this task in an intuitive manner. Help them by using examples to demonstrate that shape does affect volume.

2. Require students who are somewhat advanced to make use of volume formulas as they attempt to determine the best shape.

Chapter 75

The Three M's (Mean, Median, and Mode)

Grades 3–8

☒ Total group activity
☐ Cooperative activity
☒ Independent activity
☐ Concrete/manipulative activity
☒ Visual/pictorial activity
☒ Abstract procedure

Why Do It:

Students will enhance their understandings of statistical graphs and the three measures of central tendency, mean, median, and mode.

You Will Need:

This activity requires graph paper for each student, as well as Post-it notes or 2-inch squares cut out of construction paper, crayons or markers, and pencils.

How To Do It:

1. Begin by deciding what the students will graph. Gather some information about them that can be counted, such as the number of books they have read for the week or month, the number of hours of TV they watch in a day or week (rounded to the nearest hour), or the number of letters in their first names.

2. On the chalkboard a list with two columns will be formed when each student writes his or her name, and next to the number associated with the items they are measuring. For example, Tessa has read 2 books this month, Tessa watches 2 hours of TV each week, or Tessa has 5 letters in her first name.

Name	Number of Books
Tessa	2
Oscar	3
Tara	1

3. Each student will then make a bar graph of the class results on their graph paper. The bars should be lined up from the shortest to the longest. Because each student represents one bar on the graph, label the horizontal axis with names of students. A bar is made up of squares representing 1 unit on the vertical axis; if Tessa has read 2 books this month, therefore, a stack of 2 squares forms the bar with her name on it. You can demonstrate this by stacking Post-it notes on the board above Tessa's name to make a bar (see Example 1).

4. After all the names and bars are represented on the graph, instruct the students to find the mean of the data. To demonstrate how to find the mean (or average) number of books read in a month, move the Post-it notes from the taller bars to fill in the shorter bars, until all the bars have the same number of Post-it notes, or as close as possible. Students can do this on their own graph by crossing off squares they are removing and placing them in their new location in a different color (see Example 2).

5. The students will next find the median by going back to their original graphs, which have the bars arranged from smallest to largest. To find the median, they are to count the number of bars. If there is an odd number of bars, the middle bar will designate the median number of books. In Example 1, for example, there are 9 bars, so the middle, or fifth bar is the median. Therefore the median is 2 books. If there is an even number of bars, the two bars in the middle will be added up, and that sum then cut in half. Demonstrate this by removing the middle two bars of Post-it notes and cutting that set of Post-its in half, as shown in Example 3.

6. Lastly, the students will find the mode of the data. To do this, they need to look at the height of each bar and find the height that occurs the most. In Example 1, the height of the bar that occurs the most is 2. Therefore, the mode for the data is 2 books.

Examples:

1. The bar graph below represents the number of books each person has read in a month.

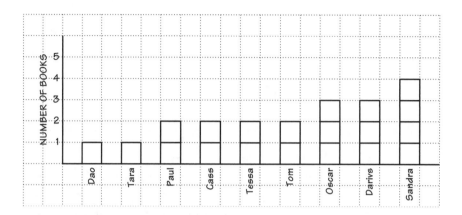

2. In the figure below, the squares that have been moved and the resulting diagram indicate that the mean is between 2 and 3. This means that, on average, a student in this class reads between 2 and 3 books a month. Students can show on their graph the moving of the squares with arrows as shown below, and then draw the resulting graph.

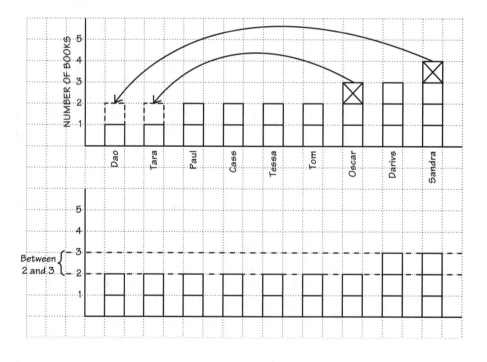

Investigations and Problem Solving

3. Taking the middle two bars from the graph below yields a total of 4 squares. Cutting this set in half reveals that the median is 2 for this data set.

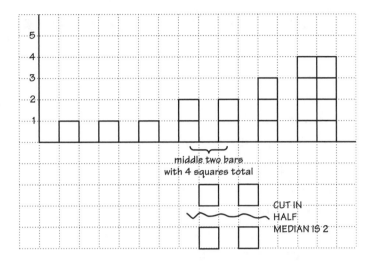

Extensions:

1. Have students do this activity for a different set of data.

2. Discuss as a class the difference between mean, median, and mode. For example, ask students whether they think it is possible to have more than one mode, or to get a mean, median, and mode that are the same number. Have students predict what the graph might look like.

Chapter 76

Post-it Statistics

Grades 3–8

☒ Total group activity
☒ Cooperative activity
☐ Independent activity
☒ Concrete/manipulative
☒ Visual/pictorial activity
☒ Abstract procedure

Why Do It:

This activity enhances students' comprehension of statistical data gathering, related graphing techniques, and probability expectations.

You Will Need:

Students will require Post-it notes (or 2-inch paper squares and masking tape), one piece of string cut less than 12 inches for each pair of students (all pieces must be of the same length), rulers, and pencils.

How To Do It:

Divide the students into groups of two. Give each pair the previously cut string. Each pair of students should look at their string and estimate its length to the nearest inch and write their estimated lengths on Post-its. Students should then carefully measure the string and discuss whether their estimates were too long, too short, or just right; they should then complete the two problems below.

1. First draw a number line on the chalkboard labeled with inch measurements. Then ask each pair of students to put their Post-its above their estimated

measurements for the piece of string. Using the graph, have the students tell which estimated measurement is the *mode* (the one with the greatest number of Post-its) of the data. Then have students determine the *range* (the difference between greatest and least estimate) for the data.

2. Next, have the students rearrange all of the Post-it estimates in a straight line from the smallest to the largest. The students can then determine which measurement is the *median* (the estimate in the middle). Next, have students find the *mean* or *average* (the sum of all the estimates, divided by the number of Post-its) estimate for the class.

Example:

The following figure shows a graph made up of 14 estimates from a class of 28. The mode would be the estimate of 9 inches. The range is $12 - 6 = 6$ inches. To determine the median, we would take all the Post-its and line them up from smallest length to largest length. Because there are fourteen Post-its, the average of the middle two will be the median. The median for this group would be found by taking the 7th and 8th measurement, which are both 9 inches, and add them and divide by 2. The median is 9 inches. Finally, the average would be $(6 + 7 + 7 + 8 + 8 + 8 + 9 + 9 + 9 + 9 + 10 + 10 + 11 + 12) \div 14$, which is about 8.8 inches.

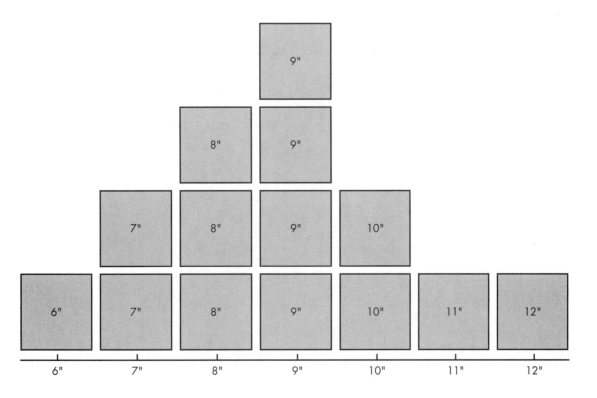

Extensions:

Give students the following problems:

1. If you had the following scores on 100-point tests, what would your *mean* or *average* score be? Your *median*? Your *mode*? Your *range*?

   | 89 | 96 | 78 | 81 | 96 |

2. It has been said that the average family has 2.5 children. Explain the meaning of this statistic.

3. What statistics would you like to know about? (For example, your statistic might involve the range in prices for a certain type of new shirt, the typical number of puppies in a litter, or the average cost for a new bicycle you'd like to have.) Write your statistical question and then search out the needed data. Ask a variety of people for help. Share your findings.

Chapter 77

A Postal Problem

Grades 4–8

☒ Total group activity
☒ Cooperative activity
☒ Independent activity
☒ Concrete/manipulative activity
☒ Visual/pictorial activity
☒ Abstract procedure

Why Do It:

Students will review concepts from geometry and apply mathematical skills, including logical thinking, and computation with a calculator, to an everyday problem-solving situation.

You Will Need:

This activity requires pencils; paper; and a calculator (recommended, but not required). Also, if some of the boxes are to be constructed, large pieces of cardboard or tagboard, scissors, and tape will be required.

How To Do It:

1. Share the following U.S. Post Office shipping problem with the students and ask how they would attempt to solve it:

> U.S. Post Office regulations note that packages to be shipped must measure a maximum of 108 inches in length plus girth. What size rectangular box, with a square end, will allow you to send the greatest volume of goods?

2. After the students have shared various ideas, they should diagram (or physically construct with tagboard and tape) one or two of the boxes they proposed and determine their volumes. When students are calculating the volumes, be certain that they understand how to relate the Length + Girth measurements to the formula Volume = Length × Width × Height (volume of a rectangular solid). It may help to refer to the Example in order to do this.

Example:

Three boxes of different dimensions, but each totaling 108 inches in length plus girth, are shown on page 299. The width and height measurements have, in each case, been derived from the initial girths (at the square ends of the boxes). The computed volumes for each are different, and students may find a calculator very helpful in finding these. Have students determine whether the square-ended box shown will yield the greatest volume, or whether another will be better. (*Hint:* The best arrangement is two cubes piled one on top of the other.)

Investigations and Problem Solving

Box Dimensions	Length x Width x Height = Volume (in Cubic Inches)
108 inches (total) − 40 inches (girth) = 68 inches (length)	68 inches × 10 inches × 10 inches = 6,800 cubic inches
108 inches (total) − 48 inches (girth) = 60 inches (length)	60 inches × 12 inches × 12 inches = 8,640 cubic inches
108 inches (total) − 80 inches (girth) = 28 inches (length)	28 inches × 20 inches × 20 inches = 11,200 cubic inches

Extensions:

1. Some students may need to physically compare the volumes. It might be useful therefore to build boxes of tagboard or cardboard to specified dimensions (such as those in the Example above). Then have students compare the volumes of the different-shaped boxes by pouring the contents of one into another; Styrofoam packing chips work well when doing this comparison.

A Postal Problem

2. Inform students that the United Parcel Service (UPS) allows boxes up to 130 inches in length plus girth. Have them determine what size rectangular box, with a square end, will allow them to send the greatest volume of goods through UPS.

3. Ask students to consider either the 108-inch limit or the 130-inch limit on girth and to determine what shape box or boxes will provide a greater volume than a square-ended rectangular box. Have them show their diagrams and calculations to demonstrate their solutions.

Chapter 78

Build the "Best" Doghouse

Grades 4–8

☒ Total group activity

☒ Cooperative activity

☒ Independent activity

☒ Concrete/manipulative activity

☒ Visual/pictorial activity

☒ Abstract procedure

Why Do It:

This activity provides a real-life investigation experience that may be solved in a variety of ways. Students will draw plans, from which they will construct their own "best" doghouses.

You Will Need:

Required for each participant are a piece of graph paper with 1-inch squares, a 4- by 8-inch piece of tagboard (old file folders may be cut up), tape, scissors, a ruler, and a pencil. A bag of rice, or other dry material, may be used when determining volume.

How To Do It:

1. Begin by posing the following problem:

> You have a new dog at home, and it is your job to build a doghouse. You have one 4- by 8-foot sheet of plywood, and from it you want to build the largest doghouse possible (having the greatest interior volume). You also decide that it will have a dirt floor, that any windows or doors must have closeable flaps, and that it will have a roof that rain will run off (see drawing below). Prior to construction, you must first draw a plan showing how you will cut the pieces from the plywood, and you will also need to construct a doghouse model using a 4- by 8-inch piece of tagboard.

2. Provide each student with a piece of graph paper and have them use their rulers and pencils to mark a 4-inch by 8-inch border. Explain that this bordered area will represent (as a scaled version) the 4- by 8-foot sheet of plywood. Then provide time to investigate where best to draw the lines so that the cut-out pieces will allow them to create the largest doghouse. Remind them that plywood does not bend and that all walls and the roof must be filled in. They may, however, splice together some sections of the doghouse, but very small slivers are not allowable.

3. When students have finished their plans, provide each with a 4- by 8-inch piece of tagboard, scissors, and tape. Have them mark their tagboard, cut out the pieces, and tape them together to form doghouse models. (*Note:* Any material excess should be taped inside the doghouse, and the student's name should also be written inside.)

4. When a number of the investigators have finished, allow them to compare and contrast the doghouse models that they built by discussing, for example, whether it was better to build a

long doghouse or a square one, or whether a tall doghouse provided more inside space (volume) than a short one. Advanced investigators may use mathematical formulas to determine the outcomes, but young learners may need to take a more direct approach in deciding which doghouse has the greatest volume. To do so, they may simply tape any doors and windows shut, turn the doghouse models upside down, fill one with rice (or other dry material), and pour from one to another until it is determined which model holds the most. A discussion noting the attributes of the best doghouse should follow.

Example:

The doghouse models shown below have both been built from 4- by 8-inch pieces of tagboard, but their shapes and volumes are quite different.

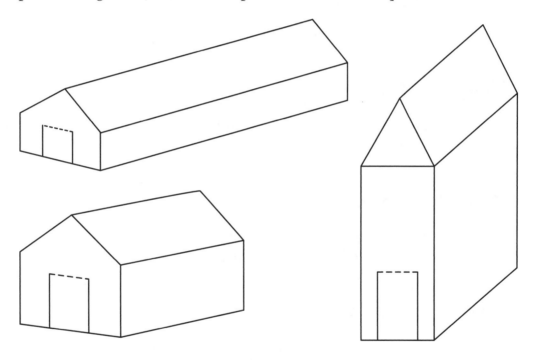

Extensions:

1. Younger students will likely need to complete this task in an intuitive manner; it may be helpful to relate it to the activity *Building the Largest Container* (p. 288). In any event, help them understand that shape does affect volume.

2. Students who are somewhat advanced should be expected to make use of area and volume formulas as they attempt to determine the optimal shape for their doghouses.

3. Advanced students might be challenged to complete this activity using a 4- by 8-foot piece of bendable material, such as aluminum.

Chapter 79

Dog Races

Grades 4–8

☒ Total group activity
☒ Cooperative activity
☐ Independent activity
☒ Concrete/manipulative activity
☒ Visual/pictorial activity
☒ Abstract procedure

Why Do It:

Students will learn about probability while enjoying a "statistical" dog race game.

You Will Need:

Dog Races requires dice, crayons or markers, and copies of the "Dog Race Chart" for each student (chart provided).

How To Do It:

This activity provides a fun way to think about some basic probabilities and what they mean. Students will toss two dice and record the number they get by adding the dice. By performing this experiment many times, students will see a pattern develop and be able to answer some interesting questions. Beginning at the top of the chart, each student is to number the dogs 1 through 13 and circle the dog that he or she thinks will win. Working in groups of four or five, students will take turns rolling the dice. After each roll, all students in the group will add the numbers on the faces of the dice. Then each student in the group will color in a square for that numbered dog on his or her own chart. This continues until one dog has won the race.

Example:

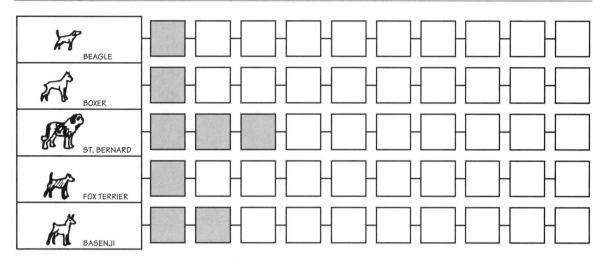

In the Example above, only a portion of the chart is shown. The following sums on the dice have been recorded, 2, 5, 4, 6, 7, 7, 5, 5. It looks like the St. Bernard is winning so far.

After the game is finished and one dog has won, or there is a tie, ask the students to answer the following questions.

1. Did you pick the winner?
2. How many times did the winning dog move forward?
3. If this race were run again, would the outcome probably be the same? (Encourage students to check by running the race again three or four times, using extra copies of the "Dog Race Chart.")
4. Which dog in this race lineup is likely to win most often? Why?
5. Are there any dogs in this race that can never win? Why?

Extensions:

1. Have the students make a chart and list the ways they can get each of the numbers 1 through 13 when using the dice. Next, ask the students to find the probabilities that each dog will win. They should write their probabilities in fraction form. For example, because there are 36 different outcomes for the sum of the numbers on two dice, then the probability that a St. Bernard would win is 4/36 or 1/9.
2. Use some different types of dice, like an octahedron (8-sided die). These can often be purchased at a school supply store, or online. Extend the "Dog Race Chart" to have some more dogs racing (for octahedron dice you would have to have 3 more dogs on the chart). Have students repeat the game above and repeat Extension 1.

3. In a real dog race, which of these dogs would be likely to win? Which might come in second, third, and so on? (You might find out about the different breeds of dogs at your library or from an expert who raises dogs.)

Dog Race Chart

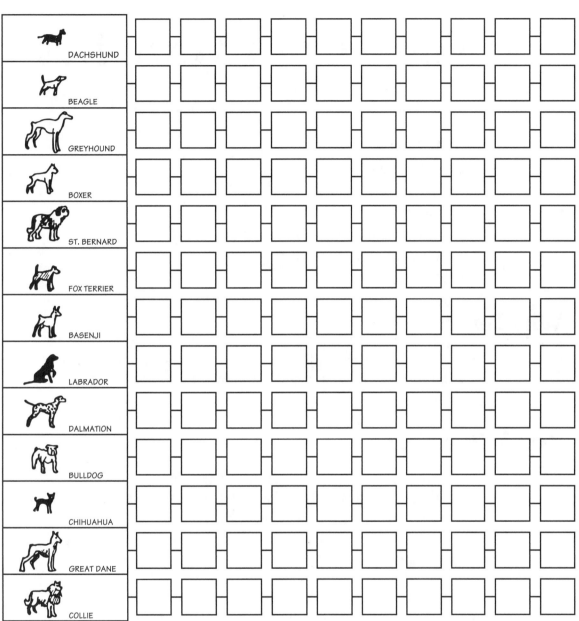

Chapter 80

Four-Coin Statistics

Grades 4–8

☒ Total group activity

☒ Cooperative activity

☐ Independent activity

☒ Concrete/manipulative activity

☒ Visual/pictorial activity

☒ Abstract procedure

Why Do It:

Students gain understanding of statistical data-gathering processes and learn how such information can be used to make predictions.

You Will Need:

Four coins and a duplicated copy of the ''Four-Coin Chart'' (or students can draw their own chart) is needed for each student, along with paper and pencils.

How To Do It:

In this activity, students will discover the patterns that develop with the heads and the tails when four coins are tossed. Students will experience performing a binomial experiment (one with only two possible outcomes) used in the study of probability and statistics.

Start by explaining to the class that each student will be tossing four coins and recording the number of heads and tails. Then have the students answer the following two questions on a piece of paper.

1. When tossing four coins at once, what combinations of heads and tails can you get? Show each of these possibilities with your own coins.

2. Predict how many of each combination that you discovered in question 1 will show up with 10 tosses of all four coins.

Next, have the students perform the experiment by tossing the four coins 10 times and recording their results on a chart like the one shown below, putting check marks in the appropriate columns. At the end of their 10 tosses, students should have 10 checks in their chart.

The students should then total the number of checks in each column and record the numbers at the bottom. Students should compare the predictions they made in question 2 above with their actual findings.

4 Heads	3 Heads 1 Tail	2 Heads 2 Tails	3 Tails 1 Head	4 Tails

Now have every student record their findings on a large chart in the classroom. Discuss with the class the totals for each column. Have students make a bar graph on their papers showing the class totals for each outcome in the chart. Last, find the probability of rolling 4 heads, 3 heads and 1 tail, 2 heads and 2 tails, 1 head and 3 tails, and 4 tails, by writing the fraction of the whole each column of results represents. For example, if there are 25 students in the class and each tossed the coins 10 times, there are 250 total tosses. If 4 heads showed up 15 times, then the probability of getting 4 heads is 15/250 or 3/50. (*Note:* this is called an experimental or empirical probability, because it is based on actually performing the experiment.

Four-Coin Statistics

Extensions:

1. Have students discover the theoretical probability for tossing four coins. This is done by finding all the possible ways the coins could display heads and tails if the order of the coins was taken into consideration. For example, when tossing four coins, 3 heads and 1 tail could come up in four different ways: HHHT, HHTH, HTHH, and THHH. To make these lists, students can use a tree diagram (see below). Then, students should write the probability based on the total of 16 different arrangements they will discover. This is called a theoretical probability because it is based on the theory of what would happen in the "ideal" experiment.

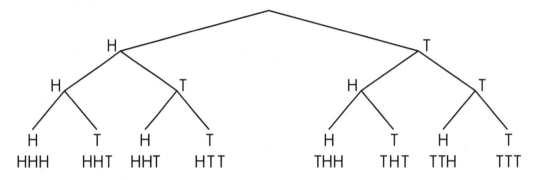

This is a tree diagram for tossing 3 coins.

2. Extend this activity further by doing the activity for tossing three coins, five coins, or more.

Four-Coin Chart

4 Heads	3 Heads 1 Tail	2 Heads 2 Tails	3 Tails 1 Head	4 Tails

Chapter 81

Tube Taping

Grades 4–8

- ☒ Total group activity
- ☒ Cooperative activity
- ☒ Independent activity
- ☒ Concrete/manipulative activity
- ☒ Visual/pictorial activity
- ☒ Abstract procedure

Why Do It:

Students will investigate a real-life problem with multiple solutions, incorporating hands-on experiences, visual mapping, and the use of formulas.

You Will Need:

A collection of paper or plastic tubes of the same diameter are needed. (Tubes of 2-inch diameter match the situation in this activity, but another size will work if the story is modified.) Also required are measuring tape, rulers, pencils, paper, and circle drawing templates (optional).

How To Do It:

Begin by posing the following problem:

> In order to raise money for a field trip, your class decided to operate a small business selling posters both on campus and by mail. When a mail order was received, the posters, which were already in tubes (with 2-inch diameters), were placed in a box that was taped shut, addressed, stamped, and sent.

One day, Julie suggested that because the posters were already in tubes, it wasn't necessary to also box them. A single tube could just be taped shut, addressed, stamped, and sent; and the same would be the case for more than one tube sent to a single address, except that multiple tubes would need to be taped together. After some discussion, everyone agreed to use Julie's idea.

Soon, however, they discovered a few problems. Jose and Tony were taping together orders for 3 posters, but each did so differently: Jose placed his posters side by side, and Tony organized his as a triangle. Susan didn't think the arrangement would make any difference, but Dan thought the triangle shape might take less tape. They decided to use a measuring tape to find out if there was a difference. What they found was that the tape needed for the side-by-side arrangement measured almost 14-3/8 inches + 1 inch overlap = 15-3/8 inches, whereas the triangle configuration was approximately 12-3/8 inches + 1 inch overlap = 13-3/8 inches.

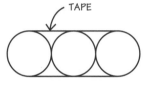

(SIDE BY SIDE WE NEED
15-3/8 INCHES OF TAPE.)

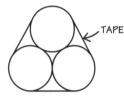

(AS A TRIANGLE WE NEED
13-3/8 INCHES OF TAPE.)

Because most of the mail orders were for more than one poster, the class decided that they needed to know the best way or ways to arrange different numbers of tubes for taping. For instance, 4 tubes (see Example) might be arranged side by side, as a square, or as a rhombus. Which arrangement or arrangements would require the least tape? Because orders of up to 7 posters had been received, the class needs to know the best arrangements for 1 to 7 tubes taped together.

Next, have students form groups of four and have each group work on finding the different ways to arrange 1 to 7 tubes. Also, the groups should find the arrangement for each number of tubes that uses the least amount of tape and is therefore the most cost-effective way to send the packages.

Example:

Shown below are possible arrangements for 4 to 7 tubes. Ask students to determine which is the best arrangement. (*Note:* See Extension 2 for a more in-depth explanation.)

4 TUBES

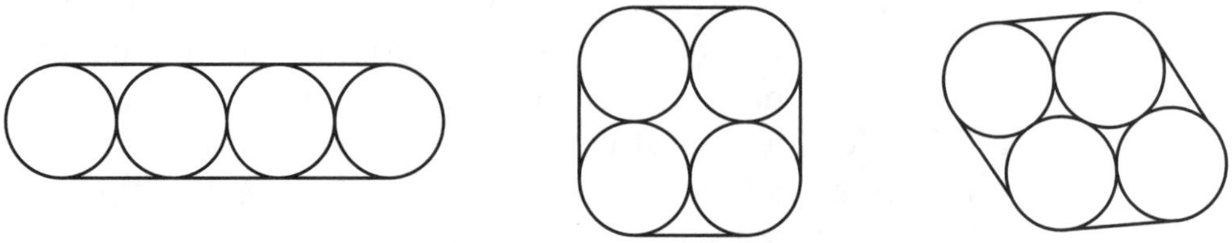

5 TUBES

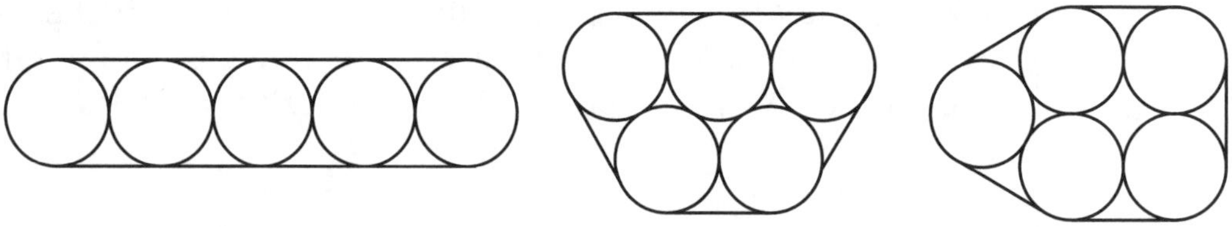

6 TUBES

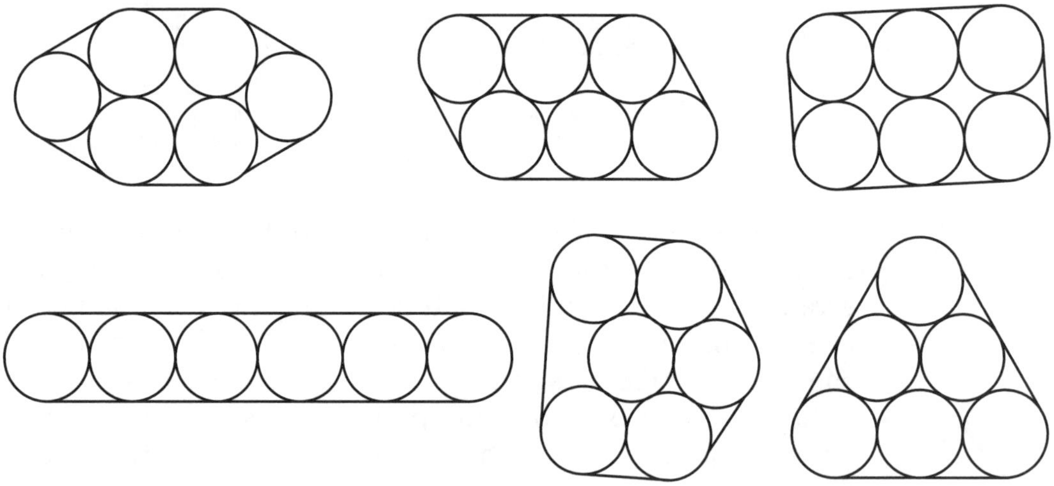

7 TUBES

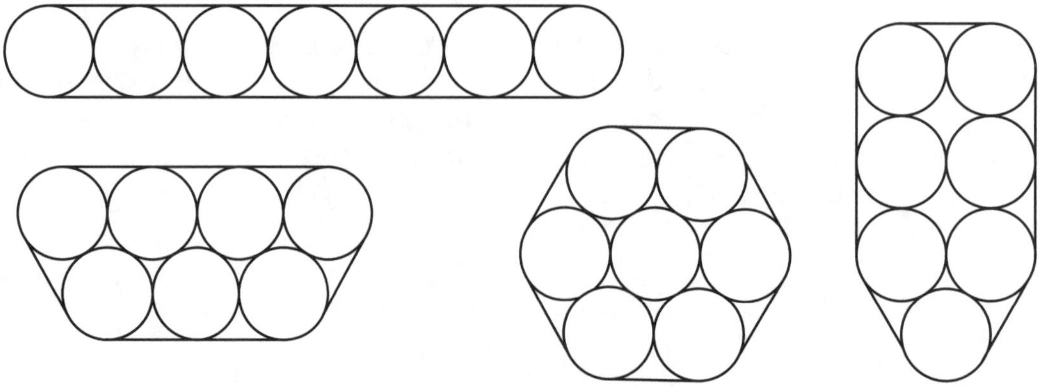

Extensions:

1. Younger students will probably need to complete the designated task by physically taping tubes together and then measuring each configuration. In this manner they will gain intuitive understandings about the best organization patterns.

2. Students who have had some experience working with basic measurement and geometry diagrams and formulas should use these to derive mappings and generalizations, such as those noted below:

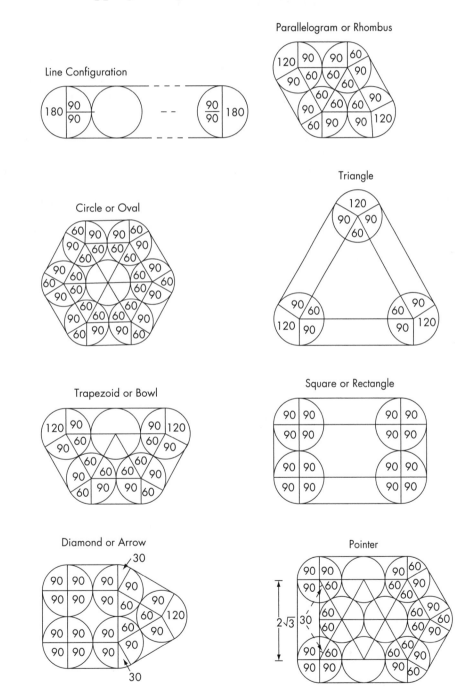

When considering tubes with 2-inch diameters, participants should conclude:

- The tape will curve around the tubes and straighten out between the tubes.
- No matter how the tubes are arranged, the curved part will always $= 2\pi$, the circumference of the circle $= 2\pi r$, and the radius $= 1$ inch.
- The straight length of tape between any 2 tubes will always be the radius of 1 tube plus the radius of the other, which in this situation is 2 inches. (The exception is the Pointer configuration.)

3. Advanced students can be challenged to consider costs, in the same manner that any small business should. They can, for example, consider the potential cost just for the tape needed to ship 100 (or more) tubes with posters in them; a variety of questions should then be posed and tentatively answered, including: How many inches of strapping tape are on a roll? What does a roll cost? How many rolls will be needed? Are the tubes to be taped at 1, 2, or 3 locations, and do the ends of the tubes need to be taped shut? About how many tubes will be sent out taped in 2s, 3s, and so on? Students can also address such concerns as the cost of postage, and how much they must charge for posters to make a reasonable profit.

Chapter 82

Height with a Hypsometer

Grades 4–8

☒ Total group activity
☒ Cooperative activity
☒ Independent activity
☒ Concrete/manipulative activity
☒ Visual/pictorial activity
☒ Abstract procedure

Why Do It:

Students will apply geometry and measurement concepts in a manner similar to that used by surveyors.

You Will Need:

Each participant requires the following: one sheet of graph paper; a piece of stiff cardboard or tagboard about 10 by 12 inches; 20 inches of string; a heavy washer or other weight; a "fat" plastic straw; tape; and glue.

How To Do It:

1. Define a *hypsometer*—an instrument that determines the height of a tree by triangulation—for your students. Each student will make his or her own hypsometer. Show them how to make one, first by gluing a sheet of graph paper on stiff cardboard and taping a plastic straw along the upper edge of the cardboard and graph paper, as shown in the Example. Have students label the right side of the straw, which is in the top right-hand corner of the graph paper, the letter A,

and the bottom right corner of the graph paper D. They will then hang a weight on a piece of string from point A. They are now ready to measure the height of objects using similar right triangles. For example, if the string swings back as shown in the figure, the point where the string hits the bottom of the graph paper is called E. The right triangle ADE is formed. As the student uses the hypsometer to look up at the tree, another triangle is formed, right triangle AJK.

2. Have students take their completed hypsometers outdoors and use them to determine the height of a tree (or building). To do so, they measure (or "pace off") the distance from the tree—perhaps 10 yards. Now each student is to hold his or her hypsometer and look through the straw until they can see the top of the tree. The weighted string will hang perpendicular to the ground. Using a finger to clamp the string in place on the cardboard, the student has now formed a right triangle on the cardboard hypsometer, where the string is the hypotenuse (longest side).

3. By sighting the triangle AJK and clamping the string in place with his or her finger, the student has automatically created a similar triangle ADE (as well as others) on the hypsometer graph paper (see Example). Students then count off the appropriate number of unit spaces along AD to correlate with the measured distance from the base of the tree; in this case, 10 unit spaces represent 10 yards. Learners next count the number of unit spaces from point D to point E, which will represent the tree's height in yards; in this case, 3 spaces denote 3 yards. Also, be certain that students take their own heights into account, because they were probably standing and sighting from eye level when they took their hypsometer readings. For example, a person just over 2 yards (6 feet) tall would find the tree to be 3 yards + 2 yards = 5 yards tall.

Example:

In this example, AJ is 10 yards, AD is 10 units on the graph paper, and DE is 3 units on the graph paper.

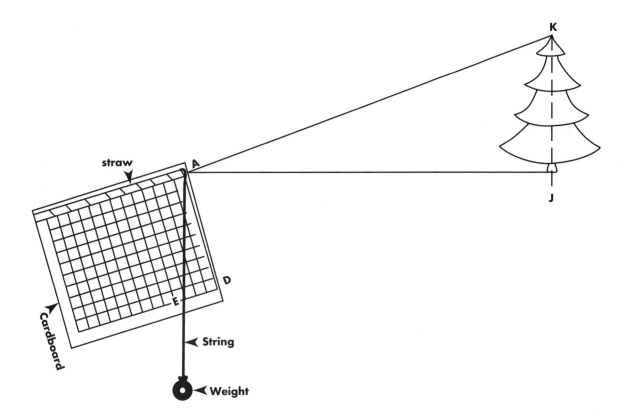

The hypsometer arrangement of similar right triangles can also be indicated through ratios. Using the same illustration above, the following proportion shows that the right triangle ADE is similar to right triangle AJK:

$$\frac{AD}{DE} = \frac{AJ}{JK}$$

As long as the sighting height is added to JK, this is an accurate way to determine the height of the tree. To solve a proportion, we can cross-multiply. For example, if AD = 10 cm, AJ = 10 yards, and DE = 3 cm, then the proportion is $10/3 = 10/x$, where x is the missing length (or height of tree). If we cross-multiply, the equation is now $10x = 30$, and solving for x we get $x = 3$ yards.

As soon as the students have grasped the concepts relating to measurement using similar right triangles, suggest that they try the procedure on objects for which the heights can readily be determined. In this way, students can check the accuracy of their sighting measurements. They might try the school flagpole (the custodian might know the actual height) or a commercial building for which the architect's plans can be reviewed. Through these investigations, students will begin to understand applications for geometry used by surveyors, forest-service personnel, and others.

Extension:

Transits and levels (which are similar to the hypsometer) are frequently used to make accurate land, architectural, or other measurements. Invite someone who uses these devices, such as a highway surveyor, an architect, or a forest-service timber cruiser, to give a class demonstration.

Chapter 83

Fairness at the County Fair

Grades 4–8
☐ Total group activity
☒ Cooperative activity
☒ Independent activity
☒ Concrete/manipulative activity
☒ Visual/pictorial activity
☒ Abstract procedure

Why Do It:

Students can practice adding wins and losses (often represented by signed numbers), finding an average, and using a simulation to solve a problem in statistics. Also, this activity reviews computation with fractions and decimals.

You Will Need:

This activity requires small paper plates, scissors, paper clips, bobby pins, masking tape, glue, spinners (reproducibles included at the end of this activity), paper, and pencils.

How To Do It:

In this activity, students will use a spinner to simulate a game at the county fair, and to decide whether the game is fair or not.

1. Here is a problem that you will read to the class. Precede the reading of the problem with a question as

to whether students have ever played games at a fair. Students should describe a game they played, and then you can ask whether they thought it was easy or hard to win the game. After this discussion, tell the students that they will learn a way to analyze a game that might be at the fair. Then, read the problem.

Joe is playing the token-toss game at the county fair. He pays $3 to throw a token, which lands on the table, as shown below. If the token lands on the $4 section, for example, Joe will receive $4, but because he already paid $3 for the game, he would be winning only $1. If it lands on the $0.50 section, he'll win $0.50, which means he actually lost $2.50. If the token lands on a line, Joe gets to throw again, until it lands inside a section.

If Joe plays this game 20 times, what would his average of wins and losses be? Is this a fair game?

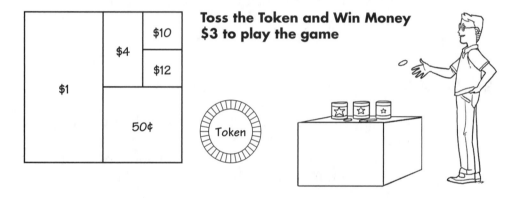

Toss the Token and Win Money
$3 to play the game

2. Before they make their own spinners, have students look at the example spinners included at the end of this activity (see reproducible spinners at end of activity) and show how each section on the circle labeled Spinner #1 is fractionally equivalent to the sections on the square board in the figure above. The $1 section makes up half of both the board and the spinner. Similarly, $0.50 is 1/4 of the board and spinner, $4 is 1/8, $10 is 1/16, and $12 is 1/16.

3. Have each student cut out Spinner #1 (photocopied from the reproducible provided) and glue it on the paper plate. He or she is to unbend the paper clip and poke it through the center hole of the spinner, looping it over above the top of plate like a hook. Then have the student bend the paper clip under the plate so that it is flat against the bottom of the plate and use the masking tape to tape the paper clip to the bottom of the plate. Next, the bobby pin is slipped over the hook in the paper clip above the plate and pushed down to lay flat on top of the

spinner. Then the hooked part of the paper clip can be bent so that when the bobby pin is spun it will not fly up and off the bent paper clip.

Making a Spinner

Cut out the paper spinner and glue it on the paper plate. Unbend the paper clip and poke it through the center hole of the spinner. Use the masking tape to tape the paper clip to the bottom of the spinner. Put the bobby pin on to function as a spinner.

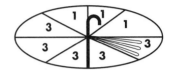

4. Explain that when players spin the spinner, they will record (on their game chart provided as part of the reproducible page) how much money they have won or lost, taking into account the $3 each paid at the start of the game. Players can record the money as a win or loss, or as a positive or negative. For example, if a student lands on the $1 section on the spinner, he or she will compute $1 won − $3 spent, which shows that he or she lost $2 for this game. He or she will then record either $2 loss or −$2 as the outcome. If the spinner lands on the $10 section, the student will compute $10 won − $3 spent, and record $7 won or +$7.

5. The players should spin the spinner 20 times and record their results on the game chart provided.

6. Require students to compute the average of their wins and losses. First have them add up the total amount of all wins and the total amount of all losses, and ask them to subtract the latter number from the former. If the total amount of all wins is greater than the total amount of all losses, the final amount will be positive (a win). If the win total is less than the loss total, however, the final amount will be negative (a loss). After they have computed their totals, have students divide by 20 to compute the average (mean) win or loss.

7. Each player now interprets his or her result in regard to whether the average showed that he or she had won or lost after 20 games. Students then can determine if it was a fair game, a game in which the county fair had the advantage, or a game in which the player had the advantage. A zero average indicates a fair game, a negative average indicates the county fair's advantage, and a positive average indicates the player's advantage.

Example:

In this example Joe has spun the spinner 6 times so far and recorded his results, of $1 win, $2 loss, $7 win, $2 loss, $2.50 loss, and $2 loss. If he adds what he has so far, he has a total loss of $0.50.

Extensions:

1. The players can perform this experiment additional times to see if the result is close to what they got the first time. Players can also combine and find the average of all outcomes. This should lead to a discussion about how the more times an experiment is performed, the more accurate the average.

2. Players can figure out the expected value for this game. Expected value is calculated by multiplying the probability of the token's landing for each section by the amount of money gained or lost, and then adding all these products together. The probability of the token's landing on a section is the fractional part of the square table the section represents. The expected value of an experiment is the average (mean) value of the outcomes for many repetitions of the experiment. The answer to this problem is $1/2(-\$2) + 1/4(-\$2.50) + 1/8(\$1) + 1/16(\$7) + 1/16(\$9) = 1/2(-2) + 1/4(-5/2) + 1/8(1) + 1/16(7) + 1/16(9) = -1 - 5/8 + 1/8 + 7/16 + 9/16 = -1 - 4/8 + 1 = -4/8 = -1/2 = -\$0.50$.

3. Change the value of each section on the square board (and on the spinner) as shown below and have the players try the simulation again. They can also compute the expected value as shown in Extension 2.

4. Change the game board to look like the one shown below. Players can then use Spinner #2 provided at the end of this activity and repeat the simulation.

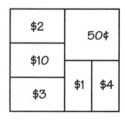

5. Players can make up their own carnival game and use the blank circles provided to make their own spinners. (The blank circles are marked in 10-degree increments.) They can then perform the simulation with their new spinners.

6. Players can also change the price to play the game from $3 to $2 or $5, and do any of the above problems again.

Investigations and Problem Solving

Fairness at the County Fair

At the County Fair, one of the carnival games is a token toss. The player pays $3 to throw a token. The token is thrown onto a table. The player will get the amount of money indicated by the section of the board on which the token lands. If it lands on a line, the player is given another throw, until it lands in a section. If you play this game 20 times, what is the average of your wins? Is this a fair game?

Game Number	Result (+) Win or (−) Loss	Game Number	Result (+) Win or (−) Loss
Game 1		Game 11	
Game 2		Game 12	
Game 3		Game 13	
Game 4		Game 14	
Game 5		Game 15	
Game 6		Game 16	
Game 7		Game 17	
Game 8		Game 18	
Game 9		Game 19	
Game 10		Game 20	

Find the mean (average). Add up all the winnings and losses (positive and negative numbers) and divide by 20.

Mean = _____

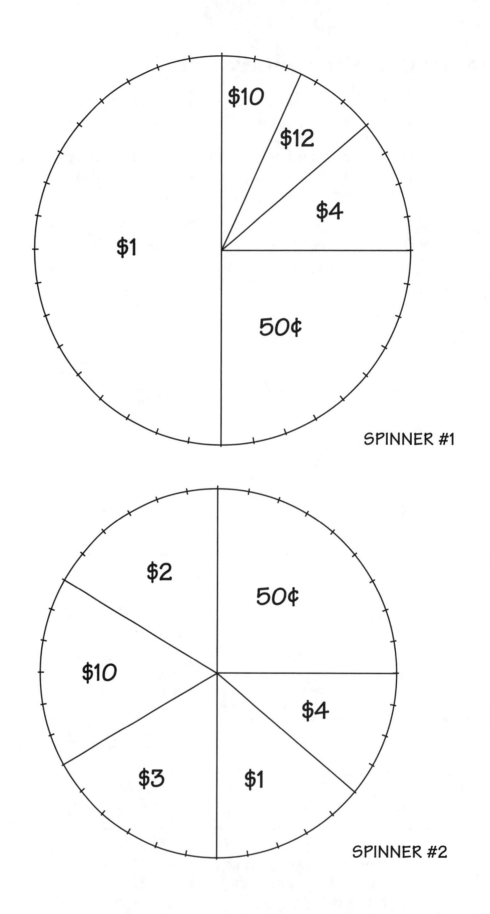

$10

$12

$4

$1

50¢

SPINNER #1

$2

50¢

$10

$4

$3

$1

SPINNER #2

Investigations and Problem Solving

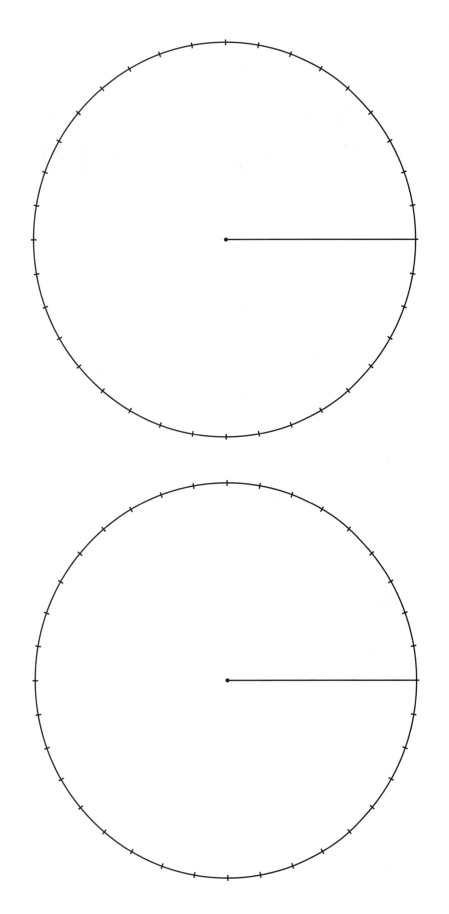

Fairness at the County Fair

Chapter 84

Winning a Prize Spelling "NUT"

Grades 4–8

- ☐ Total group activity
- ☒ Cooperative activity
- ☒ Independent activity
- ☒ Concrete/manipulative activity
- ☒ Visual/pictorial activity
- ☒ Abstract procedure

Why Do It:

This activity introduces students to the concept of a statistical simulation used in data collection for an experiment, and encourages them to analyze and interpret data.

You Will Need:

One paper bag and three white, two red, and one blue poker chips are needed for each student or group of students (you can use any different-colored objects as long as they are the same shape), as well as paper (or photocopies of the chart provided) and pencils.

How To Do It:

1. Begin by posing the following problem. (This problem is also on the reproducible handout for this activity.)

> The local health-food store is promoting eating nuts as a source of protein. At the cash register there is a large bowl containing 600 marbles of the same size. On each marble is the letter N, U, or T, printed in the ratios 3:2:1, respectively. Each time you make a purchase of a product with nuts, you can shake the bowl and pull out a marble. The clerk then gives you a ticket with the letter that shows up on the marble, and the marble is replaced into the bowl. If you collect enough tickets to spell the word "NUT," you will win a prize. About how many marbles do you think you might have to pull from the bowl before you spell the word "NUT"? (Assume you cannot see the marbles in the bowl when you draw.)

Next, discuss the idea of ratios by explaining that there are three times as many N's as there are T's, and two times as many U's as T's. This means that if there are 600 marbles, 300 have the letter N on them, 200 have the letter U, and 100 have the letter T.

2. Have the students guess as to how many marbles they will need to pull before spelling "NUT." Then have them record their guesses on their activity sheet.

3. Each student will then place 3 white poker chips, 2 red poker chips, and 1 blue poker chip into a paper bag to simulate the problem. The student will shake the bag, pull out a chip without looking in the bag, and record the letter that corresponds to the chip on his or her chart (see reproducible at the end of this activity). After the chip is returned to the bag, the process is repeated. Students are to keep shaking and drawing until all three letters N, U, and T have been written down. Then the number of chips (pulling a chip is simulating pulling a marble from the bowl) is recorded in the column on the right.

4. The students will do this experiment for a total of 10 trials, recording their results. At the end of 10 trials, ask students to compute the mean (average) of the number of marbles that must be drawn to spell the word "NUT."

5. Record the mean from each student or group on the board. Then, with the help of the class, either compute or estimate the overall mean. Have the students both compare their individual guesses with the overall mean and explain in words how the result of this experiment solves the problem.

Extensions:

1. The students can perform this experiment more times to see if their results are close to what they got the first time. This will lead to a discussion about how the more an experiment is performed, the more accurate the result (in this case, the average number of marbles required to spell "NUT").

2. Give students the following problems: (1) If there are 6,000 marbles in the bowl, how many have N, U, or T on them? (2) What about 900 marbles? (3) What about 3,000 marbles? (4) What about 1,920 marbles?
   Answers: (1) 3,000; 2,000; 1,000; (2) 450; 300; 150; (3) 1,500; 1,000; 500; (4) 960; 640; 320. (You can do this using guess and check or an algebra equation: $3x + 2x + 1x =$ total marbles.)

Winning a Prize Spelling "NUT" Activity Sheet

The local health-food store is promoting eating nuts as a source of protein. At the cash register there is a large bowl containing 600 marbles of the same size. On each marble is the letter N, U, or T, printed in the ratios 3:2:1, respectively. Each time you make a purchase of a product with nuts, you can shake the bowl and pull out a marble. The clerk then gives you a ticket with the letter that shows up on the marble, and the marble is replaced into the bowl. If you collect enough tickets to spell the word "NUT," you will win a prize. About how many marbles do you think you might have to pull from the bowl before you spell the word "NUT"?

Make a guess: I think I have to pull _____ marbles to win the prize.

Simulation:	Key:	
Shake the bag and, without looking inside, pull out an object. There are three different colors of the same object in the bag (3 of one color representing the N's, 2 of another color representing the U's, and 1 of the third color representing the T's). Look at the color and write down the corresponding letter in your chart using the key at the right. Put the object back in the bag and repeat the process until you have all the letters to spell the word "NUT." Count the number of times you drew out of the bag. Repeat 10 times.	**Color** / **Letter** N U T	

Example: Outcomes for trial = N, N, U, N, N, U, N, T; Number of marbles = 8

Trial	Outcomes	# of Marbles
1		
2		
3		
4		
5		
6		
7		
8		
9		
10		

Find the mean number of marbles: _____
Did your average number of marbles match your guess at the top of the page?

_____

Winning a Prize Spelling "NUT"

Chapter 85

Building Toothpick Bridges

Grades 4–8

- ☒ Total group activity
- ☒ Cooperative activity
- ☒ Independent activity
- ☒ Concrete/manipulative activity
- ☒ Visual/pictorial activity
- ☒ Abstract procedure

Why Do It:

This activity will engage students in a hands-on problem-solving project involving applied geometry.

You Will Need:

Each group or individual student requires one box of wooden toothpicks and any type of fast-drying glue. Paint and other decorative items are optional.

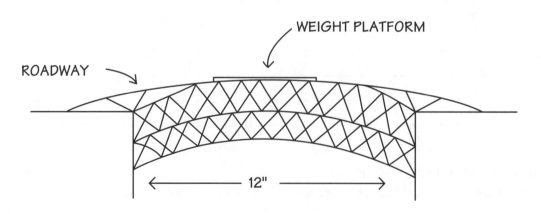

How To Do It:

Let the participants know that their next math job requires them to apply geometry by building "strong" model bridges. Their bridges may be constructed of toothpicks and glue only, they must span a 12-inch space, and they must support at least 5 pounds in the middle of the roadway (which is a surface that a toy car can travel on to cross the bridge). The bridges and related research work will be started during class time, but it is likely that it will also require some homework time. Students may work together in groups of up to four people, or they may work independently. The rules and suggestions for bridge building are as follows:

- Reference books and materials may be used to obtain information about and illustrations of different types of bridges (arch, cantilever, suspension, and so on).

- Use a maximum of 1 box of wooden toothpicks (750 count) per bridge.

- Any kind of glue may be used, but only on the joints. There must be a nonglue area completely around each toothpick.

- The support structure may be built above or below the roadway, but may not touch the floor or ceiling.

- The bridge must be at least 2 toothpicks wide, and support the weight of a toy car (at least 5 pounds).

- The length of the bridge must be exactly 12 inches between the supports, but the roadway may extend to a greater length (see diagram on page 332).

- A weight platform, measuring no more than 2 inches by 4 inches, must be built into the roadway near the center of the bridge. In this area, toothpicks may be overlapped and designed to either have weights set on it or suspended from it. The bridge must support at least 5 pounds.

- Only whole toothpicks may be used during construction. Any protruding segments must be trimmed off.

- The bridge can be painted or decorated.

Example:

This suspension bridge has its main support structure above the roadway (as contrasted with the bridge shown on page 332).

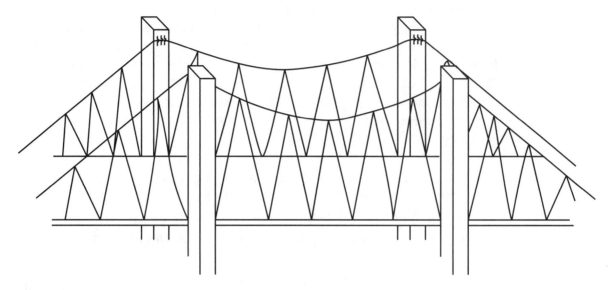

Extensions:

1. Students might draw pictures and note the special features of the following bridges (or any others they choose):

   - Drawbridge
   - Suspension bridge
   - Pontoon bridge
   - Cantilever bridge
   - Arch bridge
   - Truss bridge
   - Covered bridge

2. Students might research the history of bridges and prepare a written or verbal report.

3. Some students may be interested in comparing and contrasting important bridges. They can make a table noting, for example, the longest, highest, longest single-span, oldest covered, and most expensive bridges.

Chapter 86

A Bridge with a Bulge

Grades 6–8
☒ Total group activity
☒ Cooperative activity
☒ Independent activity
☐ Concrete/manipulative activity
☒ Visual/pictorial activity
☒ Abstract procedure

Why Do It:

This activity challenges students with an applied problem-solving experience that involves geometry, measurement, computation, and logical-thinking skills.

You Will Need:

Students will require pencils and paper. Calculators are optional but helpful.

How To Do It:

1. Share the following bridge-related situation with students and ask them to make initial estimates, draw a diagram, and discuss possible solution strategies (see *A Problem-Solving Plan*, p. 242), before attempting to solve the following problem:

A long bridge has pilings (or posts) only at the ends, and extends exactly 1 kilometer between them. The roadway on the bridge is perfectly flat, except during summer heat. On the hottest days, the roadway expands a total of 2 meters, which means that the length is 1 km + 2 m and (because there are pilings at the ends) bulges upward. The roadway has now formed an arc with its highest point in the center, with its endpoints still attached to the pilings at the same place as the original flat roadway. How high is the bulge in the road?

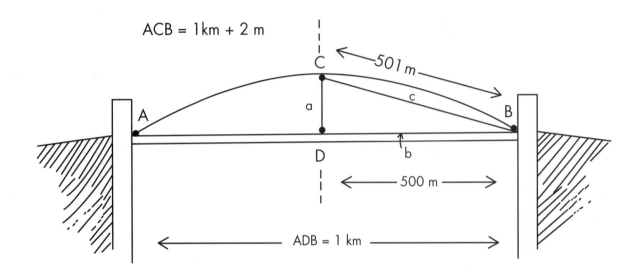

ACB = 1 km + 2 m

ADB = 1 km

2. If necessary, provide students with clues or suggestions regarding problem-solving strategies as they work on this problem. A good first strategy is to draw a diagram (with some points already labeled for easy reference, as shown above) of the original bridge and superimpose it on the bulged bridge; students should also label their bridges with reference points and known distances. This is a good time to have students make some initial estimates. For example, they can estimate the width of the gap between the original roadway and the bulged roadway (the gap labeled CD in the figure above). Because this is just an estimate, no numbers should be used, but instead they should decide whether the gap would be big enough to drive a semi-truck through, or walk under, or crawl under, or slip a hand between, or slide a sheet of paper through.

3. A discussion should follow noting that because AB = 1 kilometer, both AD and DB must be 1/2 kilometer or 500 meters. It is known that, along the bulged surface, AB = 1 kilometer + 2 meters; thus, both AC and CB = 1/2 kilometer + 1 meter, or 501 meters. It is necessary to solve for the distance CD. Further discussion should elicit some plausible solution strategies, but it may be necessary

Investigations and Problem Solving

to provide a clue by drawing a line segment CB (or AC) and asking how it might be helpful in solving this problem. Someone will almost certainly point out that because now there is a right triangle BCD, the Pythagorean Theorem ($c^2 = a^2 + b^2$) may be used. In this situation, $c = CB$ (the hypotenuse), $a = CD$ (one leg of this right triangle), and $b = DB$ (the other leg).

4. At this juncture, students should compute as follows:

$$c^2 = a^2 + b^2$$
$$(501)^2 = (a)^2 + (500)^2$$
$$251,001 = a^2 + 250,000$$
$$a^2 = 251,001 - 250,000$$
$$a^2 = 1,001$$
$$a = \sqrt{1,001}$$
$$a \approx 31.6385 \text{ meters}$$

Therefore, the bulge height is approximately 31 meters.

This solution will surprise a number of the participants, but it is approximately the correct outcome. (*Note:* The bulge curvature, when calculated, will yield a slightly different but very close result.) A number of students may need to rethink the problem or recalculate their solutions, and some may even need to build a scale model (see Extension 1 below).

Extensions:

1. Some students may need to build a scale model of this bridge situation in order to fully comprehend the activity. To do this they can glue two pieces of wood (approximately 1 inch thick) on a board such that there is exactly 1 meter (1,000 millimeters) between them. Have them cut a piece of stiff but flexible material (such as balsa wood or sheet metal) to a length of 1,002 millimeters and force it between the cross pieces. The result will be a scale model that looks and measures much like the diagram above.

2. This same problem can be done using English measurement units. In this case, the bridge would measure 1 mile long, and the expansion would be 2 feet.

3. Challenge advanced students to determine the bulge height for very small expansions, such as 1 centimeter (or 1 inch) or less.

Section Four

Logical Thinking

Throughout our lives, we use reasoning to attempt to find correct answers. The scientist seeking an explanation for some natural occurrence makes use of reasoning processes to solve his problems, as does the toolmaker trying to design a tool for a specific job or a swimmer intent on improving his technique. Logic provides the tools to guide their reasoning in the right direction. The activities and puzzles included in this section will allow students to practice both deductive logic (beginning with general principles and then proceeding to the particular case under investigation) and inductive logic (beginning with a limited number of specific facts and deriving a general conclusion).

Activities from other parts of this book will also prove helpful as students develop their logical-thinking skills. These are *Number Cutouts* (p. 22), *Reject a Digit* (p. 57), and *Number Power Walks* (p. 91) from Section One; *Chalkboard or Tabletop Spinner Games* (p. 139), *Square Scores* (p. 163), and *Here I Am* (p. 189) in Section Two; and *Sugar Cube Buildings* (p. 219), *Flexagon Creations* (p. 228), *Building the Largest Container* (p. 288), and *A Postal Problem* (p. 297) from Section Three.

Chapter 87

Stacking Oranges

Grades 2–8

☒ Total group activity
☒ Cooperative activity
☒ Independent activity
☒ Concrete/manipulative activity
☒ Visual/pictorial activity
☒ Abstract procedure

Why Do It:

Students will enhance their logical-thinking skills as they seek first hands-on and then abstract solution patterns for an everyday problem.

You Will Need:

This activity requires a bag of 35 oranges (or, in the case of allergies to oranges, balls all of the same size), and 4 pieces of 2- by 4- by 18-inch lumber (or heavy books) for the base framework.

How To Do It:

1. Tell the students that for this activity they will need to stack oranges, as grocery stores sometimes do. Ask them how they think orange stacks stay piled up without falling down. Discuss how the stacks are usually in the shape of either square- or triangular-based pyramids. Then allow the students to begin helping with the orange-stacking experiment.

2. The players might begin by analyzing patterns for square-based pyramids of stacked oranges, because these are sometimes easier to conceptualize than pyramids with triangular bases. Have them predict and then build the succeeding levels. The top (Level 1) will have only 1 orange. Challenge students to determine how many oranges will be required for the next level down (Level 2). After discussing the possibilities for Levels 3 and 4, build the structure as a class. Ask students how they might determine the number of oranges that would be needed to build an even larger base (Level 5), given that there are not enough additional oranges to build one.

3. It may be sufficient for young students to predict and build the structures for Levels 1 through 4. As they build, students in grades 2 through 5 will develop their logical-thinking skills. Older students (grades 6 through 8), however, will often logically analyze the orange-stacking progression and be able to discover a pattern and eventually a formula for determining the number of oranges at each level. Students will find that from the top down, Level 1 = 1 orange; Level 2 = 4 oranges; Level 3 = 9 oranges; Level 4 = 16 oranges; and Level 5 will require 25 oranges. Have students determine how many oranges will be needed for Levels 6, 8, 10, or even 20, instructing them to write a statement or a formula that they can use to tell how many oranges will be needed at any designated level (see Solutions).

4. When they are ready, students can be challenged with stacking oranges as triangular-based pyramids. With 35 oranges, participants will be able to predict, build, and analyze Levels 1 through 5 of the pyramid. Ask them further to determine how many oranges will be needed for Level 6, Level 10, and so on. As before, instruct them to write a statement or a formula that will find how many oranges will be needed at any designated level (see Solutions).

Example:

The students below have diagrammed the oranges needed at each level of a square-based pyramid stack. Their comments help reveal their logical thinking.

Extensions:

1. When they are finished with the orange-stacking experiments, allow participants to eat the oranges (after they wash their hands). Also, see how the oranges might be used in the same manner as the watermelons in *Watermelon Math* (p. 232), prior to their being eaten.

2. Students can represent the findings from both the square- and triangular-based orange-stacking experiments as bar graphs, and then analyze, compare, and contrast them.

3. Challenge advanced students to create orange stacks that have bases of other shapes, such as a rectangle using 8 oranges as the length and 5 oranges as the width. Learners might also be asked to find, in the case of a 7-orange hexagon base, how many oranges would be needed in the level above it, how many they would need to form a new base under it, and so on.

Solutions:

1. *Solutions for the square-based orange-stacking experiment:* Initially, participants will often notice that Level 2 has 3 more oranges than Level 1, Level 3 has 5 more than Level 2, and Level 4 has 7 more than Level 3. This realization will allow them to figure out the number of oranges needed at any level, but the required computation will be cumbersome! A more efficient method would be for the participants to recognize that all of the levels are square numbers. That is, Level $1 = 1^2 = 1$ orange; Level $2 = 2^2 = 4$ oranges; Level $3 = 3^2 = 9$ oranges, and so on.

2. *Solutions for the triangular-based orange-stacking experiment:* The hands-on stacking of oranges in triangular-based pyramids is quite easy to comprehend; however, as the following explanation notes, the abstract-level logical thinking is a bit more complex. The participants will notice that Level 2 has 2 more oranges than Level 1, Level 3 has 3 more than Level 2, and so on. Thus it can be seen that the total number of oranges at any level is equal to the number at the prior level, plus the additional oranges needed at the new level (which, for the orange stacks, is the same as the level number). For instance, the total number of oranges required at Level 4 will be 6 oranges (the total for Level 3) plus 4 oranges (which is the level number), or $6 + 4 = 10$ oranges. The following table may help clarify matters:

Level (from the Top Down)	Number of Oranges
1	1
2	$3 = 1 + 2$
3	$6 = 3 + 3$
4	$10 = 6 + 4$
5	$15 = 10 + 5$
6	$21 = 15 + 6$

Chapter 88

Tell Everything You Can

Grades 2–8

☒ Total group activity
☒ Cooperative activity
☒ Independent activity
☒ Concrete/manipulative activity
☒ Visual/pictorial activity
☒ Abstract procedure

Why Do It:

Students will investigate, compare, and contrast the logical similarities and differences of varied objects using mathematical ideas.

You Will Need:

A variety of objects (see Examples) that have at least one attribute in common are required.

How To Do It:

1. Display two mathematical items that at first glance appear to have few, if any, similarities. For instance, the square design and the clock face shown above seem

to have little in common, but logical analysis can uncover possible similarities. Help the students to see, for example, that

- Half the square is shaded, and 1/2 an hour is indicated on the clock.
- The clock face shows four quarter (or 1/4) hours, and the square is split into 1/4s.
- They both occupy approximately the same amount of space (area).
- The perimeter of the square and the circumference of the circle are "roughly" equivalent.
- Both show 360° (central angles add up to 360°) as well as four 90° quarter sections (hands of the clock at 3 o'clock is a 90° central angle)

Also spend some time discussing ways these figures are clearly different. In many instances, students will suggest logical similarities and differences that you have not recognized.

2. After one or two examples, suggest another set of mathematical objects and have the students, verbally or in written form, *Tell Everything You Can* about the objects using mathematical terms. After students have tried some of the Examples and are familiar with the process, have them make suggestions of their own for everyone to try.

Examples:

Have students attempt the following problems. (*Note:* Some possible solutions are provided.)

1. *Tell Everything You Can* about the numbers 9, 16, and 25.
2. *Tell Everything You Can* about an orange.
3. *Tell Everything You Can* about these two circles:

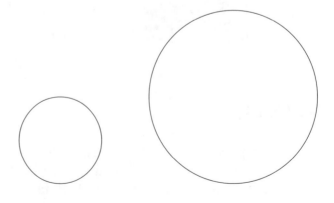

4. *Tell Everything You Can* about these two graphs:

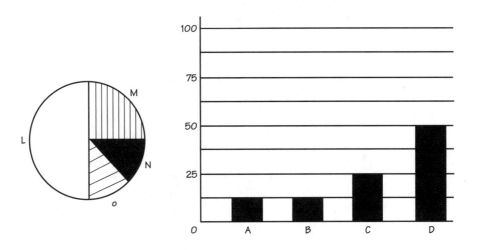

5. *Tell Everything You Can* about these two houses:

Possible Solutions:

(*Note:* Numerous other answers are possible.)

1. *9, 16, and 25:* $9 + 16 = 25$; $25 - 16 = 9$; all are square numbers, because $3^2 = 9, 4^2 = 16,$ and $5^2 = 25$.

2. *Orange:* It is almost the size of a baseball; the circumference measures as ____ inches; the peel is about 1/8 of an inch thick, and when flattened out covers about ____ square inches; there are ____ segments inside, and each is ____ (fraction) of the whole; there are ____ seeds inside.

3. *Circles:* The diameters are 1 inch and 2 inches; the circumferences are approximately 3.14 inches and 6.28 inches; at .785 square inches and 3.14 square inches, the area of the smaller circle is 1/4 that of the larger; the larger circle has about the same circumference as a Ping-Pong ball.

4. *Graphs:* Both are graphs, but one is a bar graph and the other is a circle graph. The values on the graph seem to correspond (as with L = 1/2 and D = 1/2; M = 1/4 and C = 1/4). The graph values could represent _____.

5. *Houses:* Both are "primitive" houses; both have circular bases that allow maximum floor space; the tepee is shaped like a cone, and the igloo like 1/2 of a ball or sphere; the inside volumes for the tepee and the igloo could be found with formulas if their linear measurements were known.

Logical Thinking

Chapter 89

Handshake Logic

Grades 2–8

☒ Total group activity
☒ Cooperative activity
☒ Independent activity
☒ Concrete/manipulative activity
☒ Visual/pictorial activity
☒ Abstract procedure

Why Do It:

Handshake Logic will help students understand that there are sometimes many ways to solve a single problem.

You Will Need:

Each student will need a piece of paper and a pencil.

How To Do It:

1. Introduce students to the "classic" handshake problem (see below). Have them predict possible answers and suggest how they think it might be solved.

> It is a tradition that the 9 United States supreme court justices shake hands with one another at the opening session each year. Each justice shakes hands with each of the other justices once and only once. How many handshakes result?

2. The initial predictions sometimes range from 9 to 81, and students often suggest a variety of interesting solution procedures. Because it is possible to solve this

349

problem in at least 3 or 4 different ways, have the class explore the different possibilities:

a. *Act It Out.* Ask 9 people to stand in a line. As described in the word problem, the first person in line should shake the hands of everyone else in line and then sit down, which will yield 8 handshakes. The next person in line should then shake hands with the remaining people (that would be 7) and sit down. Continue this process, being sure to record the number of handshakes, until 2 people remain in line. These 2 shake hands and record the 1 handshake between them. When totaled, the recorded handshakes equal 36.

b. *Draw a Diagram.* Using an overhead projector or the chalkboard, demonstrate how to draw 9 dots, to represent the 9 justices, in a large circle. Each student should do the same on a piece of paper. Explain that a line drawn between any two dots indicates one handshake. Instruct students to begin by choosing a dot and drawing lines connecting it to all the other dots, which will yield 8 lines, or 8 handshakes. Then have them select another dot and draw the possible lines; the outcome will be an additional 7 lines, representing 7 handshakes. Have them continue the process and count the total number of lines at the end. There will be 36.

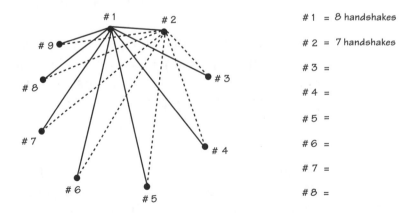

#1 = 8 handshakes

#2 = 7 handshakes

#3 =

#4 =

#5 =

#6 =

#7 =

#8 =

Logical Thinking

c. *Build a T-Table.* Tables are often useful when organizing data and looking for patterns. Have students draw the table on their papers with the left column filled in, then guide them through finishing the table. Show students that in the table, 1 person = 0 handshakes (no one to shake with), 2 people = 1 handshake, 3 people = 3 handshakes, and so on. Also, as can be seen listed below in the right column of the T-table, a related pattern evolves. Have the students fill in the right column and give them a hint as to the pattern that develops with the first few numbers. Then ask the students to finish the T-table. The last two numbers are 28 (add 21 + 7) and 36 (add 28 + 8).

Number of People	Number of Handshakes	
1	0	
		1
2	1	
		2
3	3	
		3
4	6	
		4
5	10	
		5
6	15	
		6
7	21	
		?
8	?	
		?
9	?	

d. *Use a Formula.* Many students, after trying one or more of the previous methods, may benefit from seeing how a formula can determine the same solutions that they found. The following formula, in which n = the number of justices and H = the total number of handshakes, can be used to determine the answers to the handshake problem:

$$\frac{n(n - 3)}{2} + n = H$$

Extension:

See *A Problem-Solving Plan* (p. 242) for additional techniques that can be used in conjunction with logical-thinking problems such as this one. There are online resources that can provide the teacher with more problems to solve using different problem-solving techniques. One Web site is www.abcteach.com/directory/basics/math/problem_solving.

Chapter 90

2- and 3-D Arrangements

Grades 2–8

☒ Total group activity
☒ Cooperative activity
☒ Independent activity
☒ Concrete/manipulative activity
☒ Visual/pictorial activity
☒ Abstract procedure

Why Do It:

In this activity, students will design 2-dimensional geometric arrangements and then determine which of these can be folded to make a 3-dimensional container.

You Will Need:

Several sheets of grid or graph paper, a pencil, and scissors are required for each student. (For reproducibles of 1-inch and 1-centimeter graph paper, see *Number Cutouts*, p. 22.)

How To Do It:

1. Have students begin by exploring the following 2-dimensional problem. If they wish, they may cut out graph paper squares to use in place of postage stamps. They should find as many usable arrangements as possible, keep a record of them, and compare their own findings with those of the other participants.

How many ways can you buy 6 attached stamps at the post office? Make drawings to show at least 15 different ways. Two of the 15 different ways are shown below.

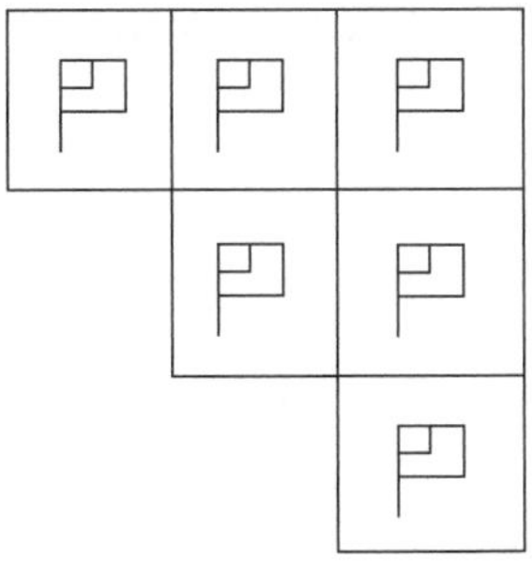

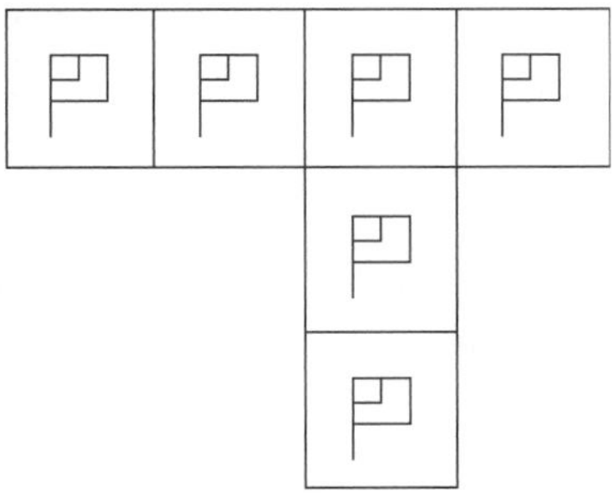

2. After exploring the numerous stamp-problem solutions, tell students to get ready to work through a related but slightly more difficult 3-dimensional problem. The 3-dimensional problem will involve the same stamp drawings, but the squares will be folded to make a closed box. With this in mind, present the following problem:

What are all possible 2-dimensional patterns, using 6 attached squares, that can be folded to form closed boxes? Make drawings of your patterns on graph paper, cut them out, and fold them along the edges to show which patterns will work. One such pattern is shown.

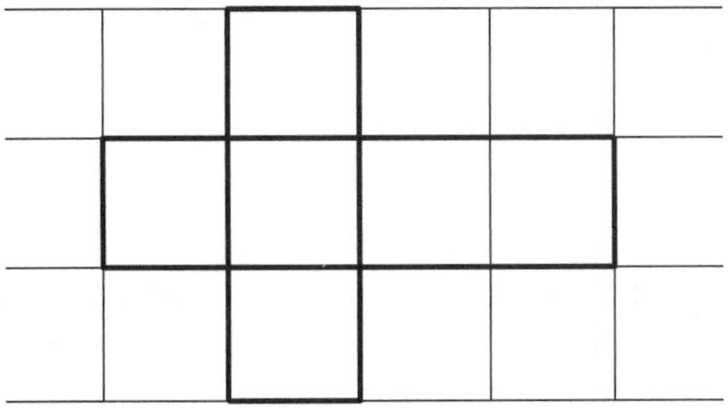

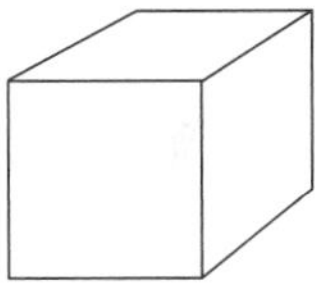

(FOLDED)

Examples:

1. On the left side of the first drawing below is a possible solution to the 2-dimensional stamp problem. The diagram on the right, however, is not feasible because some of the stamps are not fully attached along the edges.

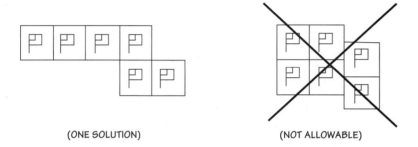

(ONE SOLUTION)   (NOT ALLOWABLE)

2. In regard to the 3-dimensional box problem, the pattern shown on the left of the next drawing can be folded into a closed box, but the pattern on the right cannot.

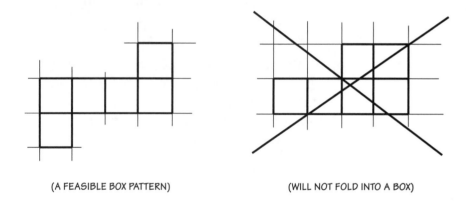

(A FEASIBLE BOX PATTERN)   (WILL NOT FOLD INTO A BOX)

Extensions:

1. Inform students that some countries use stamps shaped like triangles. Ask students to determine how many different ways 4 triangular stamps might be attached and ask them to make drawings of the possibilities. Also ask how many of these arrangements can be folded to form closed containers shaped as triangular-based pyramids, or tetrahedrons.

2. Extend the activity to include a variety of 2-dimensional patterns that, when cut out and folded, can make selected 3-dimensional figures, such as dodecahedrons or icosahedrons.

Chapter 91

Overhead Tic-Tac-Toe

Grades 2–8

☒ Total group activity
☒ Cooperative activity
☐ Independent activity
☐ Concrete/manipulative activity
☒ Visual/pictorial activity
☒ Abstract procedure

Why Do It:

Students will employ logical-thinking strategies that are used while playing a game, and will practice these skills while also learning about coordinate graphing.

You Will Need:

This activity requires an overhead projector and transparencies, or any overhead device; overhead pens; and for each group of students, graph paper and a pencil.

How To Do It:

1. Organize the players into two cooperative groups of at least four players, and designate the roles of encourager, clarifier, recorder, and speaker to four of the players in a group. The *encourager* keeps the group thinking about their next move, the *clarifier* tries to analyze the possible moves the group is thinking about, the *recorder* puts their mark on the overhead projection, and the *speaker* explains to the rest of the class the reason the group chose that spot. Using the overhead

projector, play one or two regular tic-tac-toe games to see that the designated students are properly carrying out their roles.

2. Display a Super Tic-Tac-Toe grid (see illustration below) and explain that the game only ends when all spaces have been filled with teams' marks. Points are to be awarded as follows:

6 in a row = 4 points

5 in a row = 3 points

4 in a row = 2 points

3 in a row = 1 point

LET'S PUT OUR MARK AT 4-F; THEN NEXT TURN WE CAN GO TO EITHER 5-G OR 2-D FOR A POINT.

3. Encourage each group to strategize in order to obtain the most points. Warn students that at first each group will be allowed up to 2 minutes to select and call out a location for their team's mark, but that the time will soon be shortened to 1 minute, or even 30 seconds.

4. After playing two or three Super Tic-Tac-Toe games, take time to discuss the strategies students used. For example, they may have tried to place their marks so that both ends were open, attempted to block the other team, or deliberately placed marks a certain distance apart before filling in the middle.

Extensions:

1. When students are ready, review or introduce coordinate-graphing procedures using *x*- and *y*-axis locations. Then play "Positive-Quadrant Super Tic-Tac-Toe," in which the teams' marks are

placed on the vertices (rather than in the spaces). For example, in the game below, Team O has marks at (1,2) and (1,3), and Team X has thus far placed theirs at (2,3) and (3,3). Because it is Team O's turn, they may choose to score by placing a mark at either (1,1) or (1,4); or they may choose to block Team X by marking (4,3).

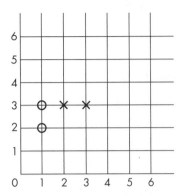

2. Play "Four-Quadrant Super Tic-Tac-Toe" in the same manner as the version described in Extension 1, except in this case both positive and negative coordinate locations must be considered. For instance, Team O has placed their marks at (−1,4) and (−3,2), and Team X has thus far marked (−2,−1) and (−2,−4).

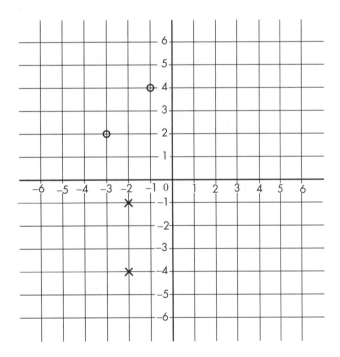

3. In advanced classrooms, allow three or four teams to play simultaneously on one large, four-quadrant grid.

Chapter 92

Magic Triangle Logic

Grades 2–8

☒ Total group activity

☒ Cooperative activity

☒ Independent activity

☐ Concrete/manipulative activity

☐ Visual/pictorial activity

☒ Abstract procedure

Why Do It:

Students will learn to logically manipulate the same numbers to achieve multiple solutions, and practice mental mathematics.

You Will Need:

The "Magic Triangle Worksheet" should be duplicated for each pair of students. Also, scissors and a pencil for each group are needed.

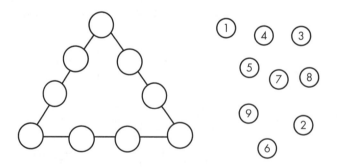

How To Do It:

A Magic Triangle is a triangle made up of 4 circles on each side, including the vertices. It is magic because the numbers 0 through 9 can be placed in the circles such that the sum of the numbers on each side is the same. Tell the students that there are many solutions to this problem, but you are not sure how many.

Give each pair of students a "Magic Triangle Worksheet," a pair of scissors, and a pencil. Instruct the students to cut out the numbers on the worksheet and tell them that they will be moving the numbers around on the large triangle until they find a solution. Then the group should record the solution they found in one of the smaller triangles at the bottom of the worksheet and indicate the solution below the triangle. (See Example below.)

Next, start the groups off by giving them the sum of 20, which is one of the solutions. This is the sum they will try to get on each edge of the triangle. The students will then work on moving the numbers around until they are placed in the triangle such that the numbers on each side of the triangle add up to 20. If students have the idea, let them work on finding other sums that will work as a solution. If students are having trouble, give them the other solution 17 as shown in the Example below and see if they can come up with the number in the triangle. Finally, challenge the students to find and record as many solutions as they can. This activity can continue for several days or even weeks.

Example:

Two Magic Triangle solutions are shown below; at least ten more are possible.

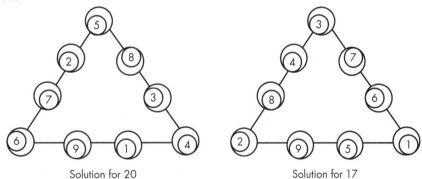

Solution for 20   Solution for 17

Extensions:

Challenge students to answer the following questions.

1. What is the greatest possible sum for each side of the triangle?
2. What is the smallest possible sum for each side of the triangle?
3. Which sum has the greatest number of different Magic Triangle combinations?
4. What other numbers will work to form Magic Triangles? (Consider 23 through 31, for example.)

Magic Triangle Worksheet

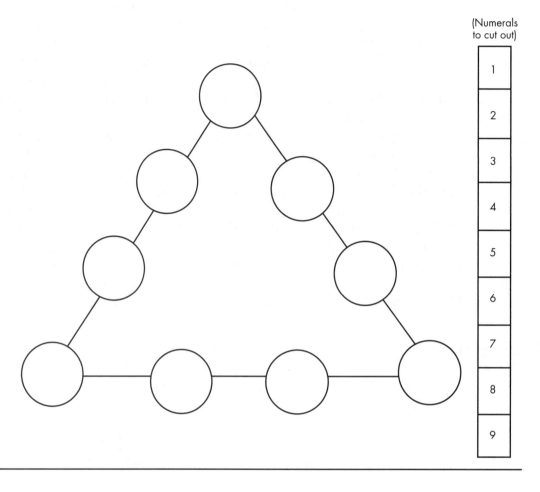

(Numerals to cut out)

1
2
3
4
5
6
7
8
9

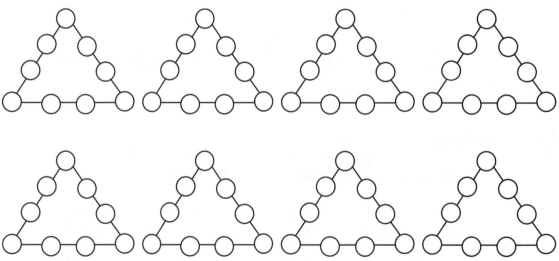

Logical Thinking

Chapter 93

Paper Clip Spinners

Grades 2–8

- ☒ Total group activity
- ☒ Cooperative activity
- ☒ Independent activity
- ☐ Concrete/manipulative activity
- ☒ Visual/pictorial activity
- ☒ Abstract procedure

Why Do It:

By creating and testing spinner surfaces, students will experience firsthand how the design of probability devices can affect their outcomes. Their designs will increase the likelihood of certain outcomes and decrease or make impossible others.

You Will Need:

Pencils, paper, and a paper clip are needed for each participant. If desired, blank spinner surfaces can be duplicated and distributed (see reproducible provided). An overhead projector or other projection device also can be used to display a variety of spinners.

How To Do It:

1. Display and discuss the possibilities for a selected spinner. Ask students whether it is possible using the spinner shown above, for instance, to spin a sum of 10 (yes, by spinning 5 and 5 again, or by spinning $3 + 3 + 4$); or to spin 11 (yes, with $5 + 3 + 3$). Discuss the different sums possible with 1, 2, 3, or more spins, as well as any sums that are impossible.

2. Have students make and test their own Paper Clip Spinners, according to the following rules:

   - Each spinner must have 3 different numbers on it.
   - The sum of the numbers can equal 12 in 3 spins, but not in 2 spins.
   - The numbers are not equally likely to occur (optional). This means that spinner sections do not all need to have the same area.

3. Give students time to work independently or in small groups as they design their spinner surfaces. Once students have completed the task, provide each of them with a paper clip to use as the spinner pointer. To do this, students lay the paper clip flat on the spinner surface so that one end overlaps with the center point. They put the point of a pencil through the end loop of the paper clip and hold the pencil on the center point with one hand. Students use the other hand to flip the paper clip, such that the paper-clip pointer randomly points to different numbers.

4. Have students try their spinners and keep records of their spins and totals. A table, included at the end of this activity, will help with record keeping. When they have finished a number of tests (four tests show in the table, but students can do more or fewer, and it will vary for each activity), have the group share and compare their findings. Students will be sharing all the possible sums they can get with 1 spin, 2 spins, and 3 spins.

Example:

Have students design a spinner that will conform to the rules that follow (two examples of such a spinner are shown here).

- Each spinner must have 4 different numbers on it.
- The sum of the numbers can equal 15 in 3 spins, but not in 2 spins.
- The numbers are not equally likely to occur.
- A number may be spun only once; if you land on the same number twice, you have to re-spin.

Students can use an enlargement of the table provided below to help determine possible outcomes and test their spinners. The number of tests they do will vary, so they can extend the table as needed.

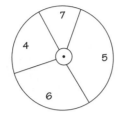
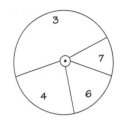

Test	Spin 1	Spin 2	Spin 3	Spin 4	Total
1					
2					
3					
4					

Extension:

Have the students experiment with the rules for spinner designs and discuss how changes affect the probable outcomes. Also have them design and test spinners with up to 10 numbers on them; they can use enlargements of the blank spinners provided here.

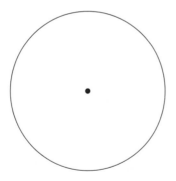
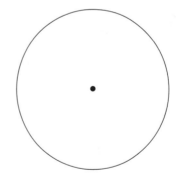

Chapter 94

Triangle Toothpick Logic

Grades 2–8

- ☒ Total group activity
- ☒ Cooperative activity
- ☒ Independent activity
- ☒ Concrete/manipulative activity
- ☒ Visual/pictorial activity
- ☒ Abstract procedure

Why Do It:

Students will enhance their logical-reasoning and pattern-seeking skills through an unusual problem-solving activity.

You Will Need:

Toothpicks or straws are required (12 for each group).

How To Do It:

In this activity, students will start by constructing a figure of triangles out of toothpicks (or straws) and will then move the designated number of toothpicks to form another figure made of triangles. Students will also be counting the number of embedded triangles in a figure.

Draw a triangular shape similar to the one here on the chalkboard or on an overhead device. Give groups of students 9 toothpicks (or straws) and have them attempt to form the figure. Have groups work on any or all of the Questions and Extensions as they manipulate the tooth-picks in various ways. Answers to these questions may be found in the Solutions, but do not give them out until students are really stuck!

Questions and Extensions:

Students are to answer each of the following questions by first using the shape shown above, except when they are directed otherwise.

1. How many small triangles did you make with the 9 original toothpicks?

2. Show how you can remove 2 toothpicks to get 3 small triangles.

3. Remove 3 toothpicks to get 1 large triangle.

4. Remove 2 toothpicks to get 2 different-size triangles.

5. Remove 6 toothpicks to get 1 triangle.

6. Remove 3 toothpicks to leave 2 triangles.

7. Start with this shape $\overline{\bigvee}\ \overline{\bigvee}\ \overline{\bigvee}$ and move just 3 toothpicks to get 5 triangles.

8. Start with the same shape as in Question 7 and move 2 toothpicks to make 4 triangles.

9. Use 12 toothpicks to make 6 congruent triangles.

Solutions:

1. 4

2. △▽△

3. (triangle figure)

4. (triangle figure)

5. △

6. △△

7. (inverted triangle figure)

8. △△▽

9. (hexagon figure)

Chapter 95

Rectangle Toothpick Logic

Grades 2–8

☒ Total group activity
☒ Cooperative activity
☒ Independent activity
☒ Concrete/manipulative activity
☒ Visual/pictorial activity
☒ Abstract procedure

Why Do It:

Students will further enhance their logical-reasoning and pattern-seeking skills through an unusual problem-solving activity.

You Will Need:

Toothpicks or straws are required (24 for each group), as well as copies of "Problem Solving with Toothpicks" (provided).

How To Do It:

In this activity, students will construct squares and rectangles out of toothpicks (or straws). Students will then be asked to move some of the toothpicks to form a figure made of squares.

Start by asking students to use 12 toothpicks or straws to make the same rectangular shape the player in the illustration has made; later, have them use 24 toothpicks or straws, as required by the problems that follow. Students will then solve as many of the problems as they can. Some problems will require more thinking than did those from *Triangle Toothpick Logic* (p. 364.)

Problem Solving with Toothpicks

Directions:

To begin, use toothpicks or straws to make the same rectangular shape as shown in the problem. Then try to solve each problem and, if successful, use a pencil to sketch your solution on paper. Notice that some of the problems require more toothpicks and become increasingly more difficult.

Questions:

Problem Solving with Squares (12 Toothpicks Needed):

1. Remove 2 toothpicks to leave 3 squares.

2. Remove 2 to leave 2 squares.

3. Remove 4 to leave 2 squares.

4. Remove 4 to leave 1 square.

5. Remove 1 to leave 3 squares.

6. Move 4 to make 3 squares.

7. Move 3 to make 3 squares.

8. Use 12 toothpicks to make 6 squares.

9. Use 12 to make 6 congruent squares.

Perimeter and Area:

10. Make a rectangle with a perimeter of 10.

11. Make a rectangle with an area of 4.

12. Make a rectangle with a perimeter of 12 and an area of 5.

Problem Solving with Toothpicks (continued)

Advanced Toothpick Problems (24 Toothpicks Needed):

13. Remove 5 toothpicks to leave 5 squares.

14. Remove 5 to leave 3 squares.

15. Remove 7 to leave 2 squares.

16. Remove 2 to leave 6 squares.

17. Remove 3 to leave 2 squares.

18. Find the minimum number of toothpicks that can be removed to leave no squares.

19. Remove 3 to leave 4 squares.

20. Remove 2 to leave 4 squares.

21. Move 3 to make 6 squares.

22. Move 4 to make 6 squares.

23. Move 2 to make 4 squares.

24. Move 1 to make a perfect square.

25. Move 4 to make 3 squares.

26. Move 3 to make 3 squares.

27. Move 4 to make 2 squares.

Logical Thinking

Solutions:

1.

2.

3.

4.

5.

6.

7.

8.

9.

10. OR

11. OR

12.

13.

14.

15.

16.

17.

18. (4)

19.

20. OR OR

21. and 1

22.

23.

24. 4

25. OR

26.

27.

Chapter 96

What Graph Is This?

Grades 2–8

☒ Total group activity
☒ Cooperative activity
☒ Independent activity
☐ Concrete/manipulative activity
☒ Visual/pictorial activity
☒ Abstract procedure

Why Do It:

Students will be introduced to different types of graphs and will investigate the meanings of these graphs from a logical perspective. Student can also find statistics related to the graphs, such as range, mean, median, and mode.

You Will Need:

This activity requires various graphs, such as those provided, that focus on everyday events or other familiar situations.

How To Do It:

Display a copy of a graph minus its title and other information that would identify exactly what it portrays. Together with students, note the type of graph, the structure of the graphed information, and whether there are numbers or other information along its axes. Then ask, ''What could this be a graph of?'' For instance, the following line graph notes an activity in which many people participate every day. Information on the axes shows that water and time are involved. (The solution to this graph is given later.)

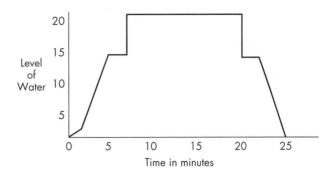

Examples:

Have students attempt the following (solutions in parentheses):

1. Which graph could represent a man riding on a Ferris wheel? (*b*)

a)

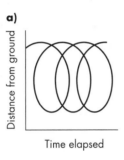

b)

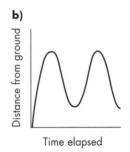

c)

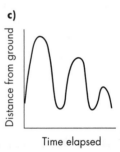

2. Which graph could represent a child climbing a slide and then sliding down? (*a*)

a)

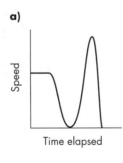

b)

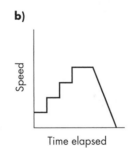

c)

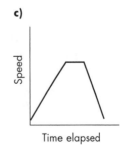

3. Which graph could represent a train pulling into a station and letting off its passengers? (c)

a)

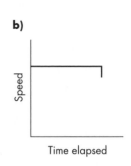

b)

c)

4. Which graph could represent a child swinging on a swing? (b)

a)

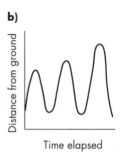

b)

c)

More Examples To Try:

Have students attempt the following problems, some possible solutions for which are provided at the end of this activity.

5. Is this a line graph, a circle graph, a bar graph, or a pictograph? What information is given on this graph? What do you think this might be a graph of?

6. In this pictograph, † = 5 students. Can you tell what this graph shows?

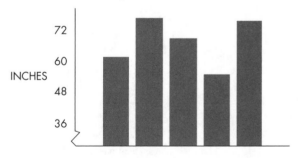

Logical Thinking

7. Can you describe what this stem-and-leaf plot shows? (*answers will vary*)

STEM-AND-LEAF PLOT

Key : 6/7 = 67

```
6 | 7  8
7 | 4  5  5  8  8  8
8 | 1  3  5  5  8  8  9
9 | 0  0  0  2  4
```

a. Find the smallest data entry and the largest data entry.

b. Find the range of data scores (largest minus the smallest).

c. Find mean, median, and mode for the data.

8. Can you describe what this dot plot is showing? (*answers will vary*)

DOT PLOT

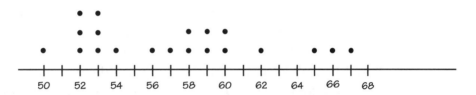

a. Find the smallest data entry and the largest data entry.

b. Find the range of data scores (largest minus the smallest).

c. Find mean, median, and mode for the data.

Extensions:

1. The graphs used in other portions of this book may prove helpful in this activity. Of interest might be those developed in conjunction with the activities *Tired Hands* (p. 278), *Four-Coin Statistics* (p. 308), and *Tell Everything You Can* (p. 345).

2. Have students explore media sources for graphs. They might, for example, bring line, bar, and circle graphs; pictographs; dot plots; or stem-and-leaf plots from newspapers or other sources to discuss with the class. (*Note:* these types of graphs may also correspond to learners' studies in science, social studies, and other subjects.)

3. Students can construct their own graphs and try them out on the other group members (and the teacher). Have them see if they can portray the same information in several different types of graphs, such as a bar graph and a circle graph.

What Graph Is This?

Possible Solutions:

1. The situation involving water and time from the How To Do It section is a line graph portraying someone taking a bath in a bathtub. During the first minute, only the hot-water faucet is turned on; then the hot and cold water is regulated and run for another 3 minutes to fill the tub; for 2 minutes the water is level (constant height); then the bather gets into the tub, raising the level (height) again, and takes a bath for 12 minutes; the bather steps out of the tub and towels off for 3 minutes; and finally, the drain is opened, and it takes 2 minutes for all the water to drain.

2. Example 5 displays a bar graph of the heights (in inches) of 5 people.

3. Example 6 shows a pictograph indicating the number of students in particular classes. Ms. Johnson has 25 students, Mr. Evans has 30, Mr. Romero has 15, and Mr. Smith has 20.

4. Example 7 could depict the ages of residents in a retirement home. (a) smallest = 67, largest = 94; (b) 27; (c) mean = 82.4, median = 84, mode = 78 and 90.

5. Example 8 could show the weights in pounds of 20 first-grade students, or the number of audience members at different youth basketball games: (a) smallest = 50, largest = 67; (b) 17; (c) mean = 57.3, median = 57.5, mode = 52 and 53.

Chapter 97

Fold-and-Punch Patterns

Grades 2–8

☒ Total group activity
☒ Cooperative activity
☒ Independent activity
☒ Concrete/manipulative activity
☒ Visual/pictorial activity
☒ Abstract procedure

Why Do It:

Students will investigate sequential patterns, determine their relationships, and learn to predict further outcomes.

You Will Need:

Each participant or cooperative group requires four or five sheets of paper (such as discarded computer paper), as well as several paper hole punches. Also, or have students sketch their own record-keeping charts.

How To Do It:

1. Begin by having the participants consider the thicknesses that result from increasingly folding paper. (The thickness of one piece of paper is measured as 1.) Ask how many thicknesses of paper would result from zero folds. Then have them fold their pieces of paper in half (Figure A below) and report how many thicknesses they have. They then fold it in half again (Figure B) and determine the new number of thicknesses. After they have folded their pieces of paper a third time, students are to make a chart to show the number of thicknesses for each fold, up to 10 folds

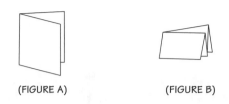

(FIGURE A)   (FIGURE B)

(see below). Finally, have each student write a relationship statement (or a rule or formula) that will express the number of thicknesses for a given number of folds. (A rule for determining thicknesses, as well as a complete chart, appear in the Solutions.)

PAPER FOLDS AND THICKNESSES

Folds	0	1	2	3	4	5	6	7	8	9	10
Thicknesses	1	2	4	8							
Rule Applied											

2. If and when the students fully comprehend the relationship between paper folds and thicknesses, have them consider the slightly more difficult Fold-and-Punch Patterns. Ask them, "If you have a sheet of paper with zero folds and you punch it once, how many holes will result? If you fold it in half and punch once again (remember that your paper already had 1 hole from punching it with no folds), what will the total number of holes be?" Then have them make a sketch (Figure C) of this situation, as well as a sketch of the number of holes that would be in the paper if each student were to fold his or her paper in half again and make one more punch (Figure D). Finally, have them both make a chart to show the total number of holes for each fold and attempt to write a rule or formula that will tell the total number of holes punched for any given number of folds. (See the Solutions for a completed example of this chart as well as the rule.)

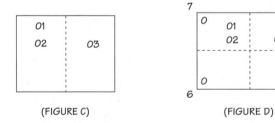

(FIGURE C)   (FIGURE D)

PAPER FOLDED IN HALVES AND HOLES PUNCHED

Folds	0	1	2	3	4	5	6	7	8
Holes Added	0	2	4	8					
Total Holes	1	3	7	15					

3. Allow the participants to note their findings and state, in their own words, the patterns found.

Extension:

After the students have solved the problems noted above, have them investigate what will happen when the paper is folded in thirds instead of halves. Have them predict and sketch the expected outcomes, punch holes, record their findings on a chart (see below), and try to derive a rule. (The completed chart, as well as the rule, appear in the Solutions.)

PAPER FOLDED IN THIRDS AND HOLES PUNCHED

Folds	0	1	2	3	4	5	6
Holes Added	0	3	9	27			
Total Holes	1	4	13	40			

Solutions:

PAPER FOLDS AND THICKNESSES

Folds	0	1	2	3	4	5	6	7	8	9	10
Thicknesses	1	2	4	8	16	32	64	128	256	512	1,024
Rule Applied	2^0	2^1	2^2	2^3	2^4	2^5	2^6	2^7	2^8	2^9	2^{10}

Rule: The number of paper thicknesses $= 2^n$, where $n =$ number of folds.

PAPER FOLDED IN HALVES AND HOLES PUNCHED

Folds	0	1	2	3	4	5	6	7	8
Holes Added	0	2	4	8	16	32	64	128	256
Total Holes	1	3	7	15	31	63	127	255	511

Rule: The holes added $= 2^n$, where $n =$ number of folds $+ 1$.

PAPER FOLDED IN THIRDS AND HOLES PUNCHED

Folds	0	1	2	3	4	5	6
Holes Added	0	3	9	27	81	243	729
Total Holes	1	4	13	40	121	364	1,093

Rule: The holes added $= 3^n$, where $n =$ number of folds $+ 1$.

Chapter 98

Coordinate Clues

Grades 2–8

☒ Total group activity

☒ Cooperative activity

☒ Independent activity

☐ Concrete/manipulative activity

☒ Visual/pictorial activity

☒ Abstract procedure

Why Do It:

This activity actively introduces students to, or reinforces their knowledge of, coordinate graphing.

You Will Need:

Coordinate Clues requires index cards or paper in three different colors (such as green, pink, and white); a marking pen; and masking tape. The amount of paper will vary depending on the number of locations on the wall and the number of clues.

How To Do It:

Write large-size numerals 0 through 9 on the green cards. Repeat this process on the pink cards. Tape these cards to two adjacent classroom walls so that they can be used to designate coordinate locations. For instance, (0,0) will be in a corner and (5,5) should be near the center of the room. Next, write sequential clues that direct students to coordinate points throughout the room on the white cards. Then hide all white cards but the first one at their appropriate coordinate locations. Put them in various spots throughout the classroom, such as under a book or taped to the bottom of a table. Put the first card at (0,0); it might read, "You are at (green 0, pink 0). For your next clue, go to coordinate (green 6, pink 3). There is a surprise waiting IF you can find your way to the end of the coordinate trail." The students may work independently or in small groups as they seek each clue and then proceed to the next. At the end of the coordinate-clue trail, each student might receive a simple surprise, such as a paper badge labeled Coordinate Clue Expert, or a coupon excusing him or her from one math problem.

Example:

The players shown below have found some of the coordinate-clue locations and the messages at each.

Extensions:

1. Play "Four-Quadrant Coordinate Clue." To do so, the (0,0) location will need to be at the center of the classroom (or playground), and the messages placed at positive or negative coordinate-graph locations as noted below.

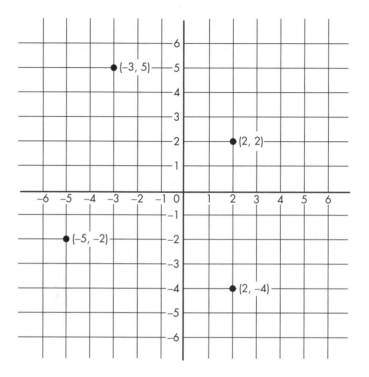

2. Advanced students might write clues and set up their own coordinate-clue trails.

3. Play "Three-Dimensional Coordinate Clue" in a manner similar to "Four-Quadrant Coordinate Clue" (see Extension 1), except that vertical coordinates and messages are also needed in this version. Some messages might be taped to light fixtures, suspended with string from the ceiling, or placed on a lower floor. (*Note:* Always think "safety first" and make sure locations are selected with this in mind.)

Logical Thinking

Chapter 99

Puzzlers with Paper

Grades 2–8

☒ Total group activity
☒ Cooperative activity
☒ Independent activity
☒ Concrete/manipulative activity
☒ Visual/pictorial activity
☒ Abstract procedure

Why Do It:

The geometry activities provided here will actively engage students in logical thinking and problem solving. The results are surprising and students will enjoy the experience.

You Will Need:

This activity requires an index card, plain paper, 5-foot strips of 3- or 4-inch-wide paper (adding machine paper works well) for each student. Each group of students should have tape, pencils, rulers, scissors, and a highlighter.

How To Do It:

1. The students will complete three paper-geometry activities. Depending on the time available, the activities may be attempted individually or all may be done during one session.

2. For the first Paper Puzzler, provide each person with an index card, a ruler, a pencil, and scissors. Each student's challenge is to cut and fold the index card such that it appears as shown below. If some of the participants are successful, allow them to share their methods with others; if not, provide a clue or two. Clues might include, for example, telling students they will make just three cuts, or that cuts will be made to a center fold.

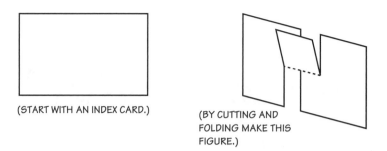

(START WITH AN INDEX CARD.)

(BY CUTTING AND FOLDING MAKE THIS FIGURE.)

3. The second Paper Puzzler challenges students to cut a piece of typing paper in order to make a loop large enough for them to step through. They will need one sheet of 8-1/2- by 11-inch typing paper, rulers, pencils, and scissors. If the players need a clue, have them make a picture frame cut on three edges of the paper (see below); then suggest that some of the additional cuts can be started at the picture frame cut, but some cannot. If the participants consider this clue carefully, they will be able to determine a logical solution (see Solutions).

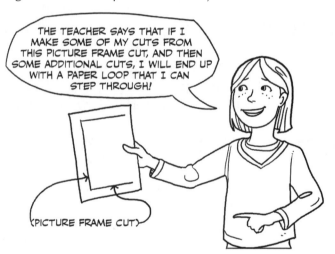

THE TEACHER SAYS THAT IF I MAKE SOME OF MY CUTS FROM THIS PICTURE FRAME CUT, AND THEN SOME ADDITIONAL CUTS, I WILL END UP WITH A PAPER LOOP THAT I CAN STEP THROUGH!

(PICTURE FRAME CUT)

4. Möbius strips provide the basis for a third type of Paper Puzzler. The Möbius strip was named after a German mathematician, August F. Möbius. It is a one-sided surface constructed by holding one end of a paper strip fixed, rotating the opposite end 180 degrees, and joining it to the first end. These strips are probably best analyzed in two phases—the 1/2 phase and the 1/3 phase.

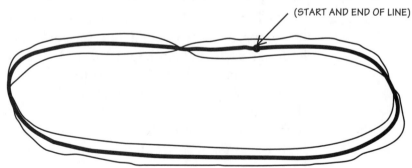

(START AND END OF LINE)

Logical Thinking

To consider a 1/2-phase Möbius strip, each participating individual or group will need a length of adding machine paper (about 5 feet), tape, a highlighter (or a pencil), and scissors. Each student first brings the ends of the paper tape together to form a large loop, and then twists one end of the loop 1/2 of a turn (180 degrees) before taping the ends together. The participants now have a loop with a twist in it, or a Möbius strip. To logically analyze their 1/2-phase Möbius strips, students should consider the following activities and questions:

a. Use a highlighter (or a pencil) to draw a line down the middle of your 1/2-phase Möbius strip. When doing so, do not lift the highlighter from the paper. What happened with the line?

b. Predict what will happen if you cut along the line that you drew down the middle of your 1/2-phase Möbius strip. Use the scissors to make that cut. What happened?

c. What do you think will happen if you cut down the middle of the "new" strip that you created? Make the cut and tell what really happened.

5. Have students consider next a 1/3-phase Möbius strip, and ask them to construct it exactly as they did the 1/2-phase Möbius strips. Have them respond, with one major difference, to problems a and b, above. This time, the line should be drawn 1/3 of the way in from an edge, and the cut should be made along that line. Something different will happen. Have students predict the outcome, and analyze whether the result is completely Möbius (a 1/2 turn or 180-degree turn) or partly non-Möbius.

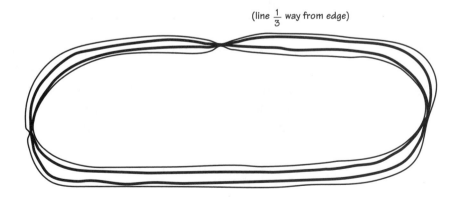

(line $\frac{1}{3}$ way from edge)

Extensions:

1. Ask the participants, "Where, in everyday life, are Möbius strips commonly used?" Or ask, "What is the advantage in using a Möbius belt, rather than a regular belt?" (See Solutions below for an answer.)

2. Have students imagine and predict the result of cutting around a Möbius strip 1/4 of the way in from an edge. Then have them do the cutting and discuss the findings.

3. Students could take a piece of paper tape, give it a full twist (360 degrees), and tape the ends together. Have them make predictions about what will happen if they cut 1/2, 1/3, or 1/4 of the way from the edge. Then students should do the cutting, repeating the experiment using three half-twists (540 degrees), and again using four half-twists (720 degrees). Have students keep a record of their findings, particularly in regard to what happened on odd-numbered twists and even-numbered twists.

Solutions:

1. *First Paper Puzzler:* Make 3 cuts as shown, fold the center tab up and crease it along the dotted line, hold the tab straight up, and give one end of the index card a 1/2 twist (180 degrees).

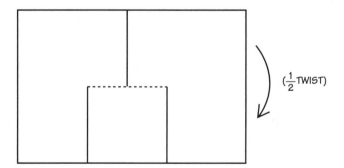

2. *Second Paper Puzzler:* Make a picture-frame-type cut along 3 edges of the typing paper (indicated on page 387). Then make 5 evenly spaces cuts in from the longest picture frame cut. Also make 6 cuts in from the non-picture-frame edge (see figure). Spread the segments into a loop and step through it.

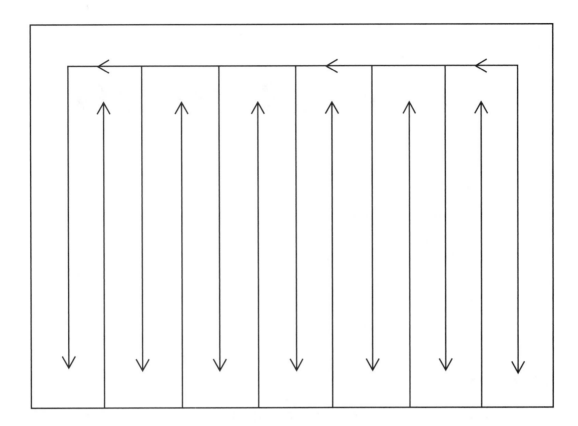

3. *1/2-Phase Möbius Strip:* (a) The line will meet itself; there is just 1 line, and therefore this Möbius strip has only side. (b) Cutting down the middle results in a single "new" strip twice as long and 1/2 as wide; further, the new strip is non-Möbius. (c) When cut down the middle, the new non-Möbius strip splits into 2 strips that are linked together.

4. *1/3-Phase Möbius Strip:* (a) The continuous line will miss itself on the "first pass," which is when you have gone all the way around the paper. But it will meet itself on the "second pass." (b) When cut 1/3 of the way in, the result will be a small, "fat" loop interlinked with a longer, "narrow" loop. The narrow loop is non-Möbius and the fat loop is Möbius. Further, the fat loop is the center of the original Möbius strip, and the narrow one is its outside edge.

5. *Extension 1:* Möbius strips are in common use as conveyor (and other) belts, because they will, theoretically, last twice as long as regular belts. The reasoning for this is that the wear is distributed evenly to all portions of a Möbius belt, whereas a regular belt wears only on one side.

Chapter 100

Create a Tessellation

Grades 4–8

☒ Total group activity

☒ Cooperative activity

☒ Independent activity

☒ Concrete/manipulative activity

☒ Visual/pictorial activity

☒ Abstract procedure

Why Do It:

This project allows students to explore regular tessellations and then create M. C. Escher–type tessellations of their own. M. C. Escher, a Dutch artist who lived from 1898 to 1972, created drawings of interlocking geometric patterns (or tessellations).

You Will Need:

Each student will require a large sheet of light-colored drawing or construction paper that is fairly stiff; a small square of tagboard (about file-folder weight) measuring 2-1/2 inches on a side; tape; pencils; scissors; rulers; and colored markers. Some examples of Escher-type tessellations or reproductions of Escher's work may also prove to be helpful. (You can find examples by going online to http://images.google.com and typing in M. C. Escher).

How To Do It:

1. A tessellation of a geometric plane is the filling of that plane with repetitions of figures in such a way that no figures overlap and there are no gaps. With this information, students are to search out and explore the many regular tessellations that are found in everyday

locations. Such everyday tessellations are most often composed of regular polygons, including squares, triangles, and hexagons (for example, ceramic tile patterns on bathroom floors, brick walls, or chain-link fences).

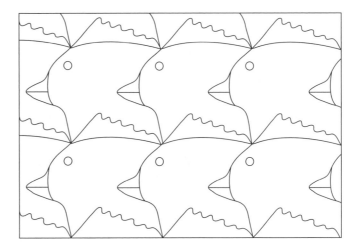

2. Next, explore some examples of the Escher-type tessellations and tell the students that they will be learning the logical procedures for developing similar tessellations of their own. When it is time to construct the tessellations, it is suggested that the class work together as they create their first tessellations; that is, the students, even though they will probably use different designs, should complete together the steps outlined in the Example.

3. After students have completed their first tessellations, engage them in a discussion of the "motion" geometry they accomplished—in this instance, the cutting out of segments and the subsequent "slide" motion to move these to their new locations. (In other instances of motion geometry, such cut-out segments might be "flipped," "turned," "stretched," or "shrunk.") This discussion, involving the logic of creating tessellations, should include such questions as What happens when you _____? and What might happen if _____? Finally, allow the participants to try out some of their ideas as they attempt the creation of more tessellations.

Create a Tessellation

Example:

Each student should follow the steps below to create his or her first tessellation.

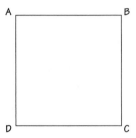

Step 1: Label your tagboard square with the vertices A, B, C, and D as shown above.

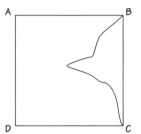

Step 2: Draw a continuous line that connects vertex B with vertex C and cut along that line to get a cut-out piece.

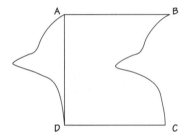

Step 3: Slide the cut cut-out piece around to the opposite side, place the straight edge BC against AD, and tape them together.

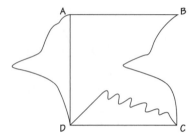

Step 4: Draw a continuous line that connects vertex D with vertex C and cut along that line.

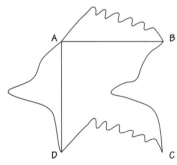

Step 5: Slide this cut-out piece around from the bottom. Place it on top, with the straight edge DC against AB, and tape them together. Your tessellation pattern is now complete.

Step 6: Place the pattern on your drawing paper and trace it. Then slide the pattern (up, down, left, or right) until it is against a matching edge and trace again. Continue until the entire drawing paper is filled with repeating patterns. You may use colored markers to emphasize your tessellation pattern.

(*Note:* See the completed bird-like tessellation on the prior page.)

Logical Thinking

Extensions:

1. The students might create tessellations to depict holidays or other important events.

2. Participants can create tessellation book covers; laminate or protect them with clear, self-stick vinyl; and mount them on their personal books or schoolbooks.

3. As a class, create a large tessellation (beginning perhaps with a 2-1/2-foot piece of cardboard) and cover an entire wall.

Chapter 101

Problem Puzzlers

Grades 4–8

☒ Total group activity
☒ Cooperative activity
☒ Independent activity
☐ Concrete/manipulative activity
☒ Visual/pictorial activity
☒ Abstract procedure

Why Do It:

Students will enhance their logical-thinking and mental-math skills while enjoying some mathematical tricks.

You Will Need:

Collect a series of *Problem Puzzlers* (see samples here). You may wish to duplicate some of the selections for individual or small group use.

How To Do It:

Problem Puzzlers may be shared verbally or in written format. In general, however, orally present the shorter problems and distribute the longer ones to the students for close scrutiny.

    A. The first group of *Problem Puzzlers*, below, should be presented orally. The answers are found in the Solutions at the end of this activity.

        1. Take 2 apples from 3 apples and what do you have?

        2. If an individual went to bed at 8:00 P.M. and set the alarm on a wind-up clock to get up at 9 o'clock in the morning, how many hours of sleep would he get?

3. Some months have 30 days, some 31; how many have 28?

4. If your doctor gave you three pills and said to take one every half hour, how long would they last?

5. There are two U.S. coins that total 55¢. One of the coins is not a nickel. What are the two coins?

6. A farmer had 17 sheep. All but 9 died. How many does the farmer have left?

7. Divide 30 by one-half and add 10. What is the answer?

8. How much dirt may be removed from a hole that is 3 feet deep, 2 feet wide, and 2 feet long?

9. There are 12 one-cent stamps in a dozen, but how many two-cent stamps are in a dozen?

10. Do they have a fourth of July in England?

11. A ribbon is 30 inches long. If you cut it with a pair of scissors into one-inch strips, how many snips would it take?

12. How long would it take a train one-mile long to pass completely through a mile-long tunnel if the train was going 60 miles per hour?

B. Some other *Problem Puzzlers* that are lengthier or that require pencil-and-paper computations are cited below. Their solutions are given at the end of the activity.

1. Suppose you have a 9- by 12-foot carpet with a 1- by 8-foot hole in the center, as shown in the drawing. Can you cut the carpet into two pieces so they will fit together to make a 10- by 10- foot carpet with no hole?

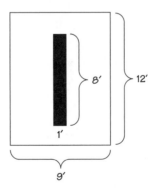

2. Johnson's cat:

> Johnson's cat went up a tree,
> Which was sixty feet and three;
> Every day she climbed eleven,
> Every night she came down seven,
> Tell me, if she did not drop,
> When her paws would reach the top.

3. *Horse trading:* There was a sheik in Arabia who had three sons. Upon his death, and the reading of the will, there came about this problem. He had 17 horses. One-half (1/2) of the horses are willed to his first son. One-third (1/3) are willed to his second son, and one-ninth (1/9) are willed to his third son. How many horses will each son receive?

4. *Rivers to cross:* There is an old story about a man who had a goat, a wolf, and a basket of cabbage. Of course, he could not leave the wolf alone with the goat, for the wolf would kill the goat. And he could not leave the goat alone with the cabbage, for the goat would eat the cabbage.

   In his travels the man came to a narrow footbridge, which he had to cross. He could take only one thing at a time across the bridge. How did he get the goat, the wolf, and the basket of cabbage across the stream safely?

5. *Jars to fill:* Mary was sent to the store to buy 2 gallons of vinegar. The storekeeper had a large barrel of vinegar, but he did not have any empty 2-gallon bottles. Looking around, he found an 8-gallon jar and a 5-gallon jar. With these 2 jars he was able to measure out exactly 2 gallons of vinegar for Mary. How?

6. *A vanishing dollar:* A farmer was driving his geese to market. He had 30 geese, and he was going to sell them at 3 for $1. "That is 33-1/3¢ a piece," he figured, "and 30 times 33-1/3¢ is $10."

   On his way to market, he passed the farm of a friend who also raised geese. The friend asked him to take his 30 geese along and sell them, too; but, since they were large and fat, he wanted them sold at 2 for $1. "That is 50¢ a piece," the farmer said his friend, "so your geese will bring 30 times 50¢, or $15."

   So the farmer decided to sell all the geese at the rate of 5 for $2. And that's exactly what he did. On his way home he gave his neighbor the $15 due him. Then he thought, "When I get home, I'll give my wife the $10 that I got for our geese." But when he looked in his pocket, he was surprised to find that he had only $9 instead of $10. He looked all over for the missing dollar, but he never did find it. What became of it?

Solutions:

A. Answers to *Problem Puzzlers* presented orally.

1. 2 apples
2. 1 hour
3. All
4. 1 hour
5. 50¢ piece + nickel
6. 9 sheep
7. 70
8. None—holes contain no dirt.
9. 12
10. Yes
11. Twenty-nine snips. The last two inches are divided by one snip.
12. Two minutes. From the time the front of the train enters the tunnel to the time the back of the train leaves the tunnel, the train must travel two miles. At 60 miles per hour, the train is going a mile a minute.

B. Answers to the longer *Problem Puzzlers* that often require pencil-and-paper computations.

1. The original carpet might be cut as shown below. Then slide the top portion to the left 1 foot and down 2 feet. The result will be a 10- by 10-foot carpet that can be sewn together or glued down.

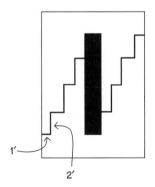 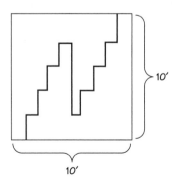

2. *Johnson's cat:* Each day, the cat went up 11 feet and came down 7. So she moved up 4 feet per day. In 13 days the cat climbed 4 × 13, or 52 feet; on the 14th day her paws reached the top, because 52 + 11 = 63.

3. *Horse trading:*

$$1/2 \times 17/1 = 8\text{-}1/2 = 9 \text{ horses for first son}$$
$$1/3 \times 17/1 = 5\text{-}2/3 = 6 \text{ horses for second son}$$
$$1/9 \times 17/1 = 1\text{-}8/9 = 2 \text{ horses for third son}$$
$$\text{Total} = 17 \text{ horses}$$

4. *Rivers to cross:* Takes goat across; returns. Takes wolf across; brings back goat. Takes cabbage across; returns. Takes goat across.

5. *Jars to fill*: Call 8-gallon jar A and 5 gallon jar B. Fill B; empty B into A. Fill B. Fill A from B. There are 2 gallons left in B.

6. A vanishing dollar:

$$\$10 + \$15 = \$25$$
$$5 \text{ for } \$2 = 40\text{¢ each}$$
$$60 \times 40\text{¢} = \$24$$

Chapter 102

Dartboard Logic

Grades 4–8

☒ Total group activity

☒ Cooperative activity

☒ Independent activity

☐ Concrete/manipulative activity

☒ Visual/pictorial activity

☒ Abstract procedure

Why Do It:

Students will learn to mentally manipulate number sums in order to determine possible game board solutions.

You Will Need:

Dartboards (a reproducible page of dartboards is provided at the end of activity, or students can sketch them) with several different number scoring patterns are required, as well as pencils.

How To Do It:

Place a dartboard in front of the classroom (or sketch one on the chalkboard or overhead projector) and play a practice game. Ask students what scores are possible when, for example, 3 darts are thrown at the target shown here. They should start by making a list of possible number combinations when throwing 3 darts and then add up the different combinations to create a list of possible scores. The students should then find the greatest and least scores to see the range of possible scores. Finally, have the students discuss what scores within the range cannot be achieved by tossing 3 darts.

Next have the students consider the same target again, but this time with 4 darts. They may work in cooperative groups or individually to determine the new possibilities; when they are finished, discuss their findings as a class. Students should also consider the possible outcomes for 2 darts, 5 darts, and 6 darts. This could lead to a discussion of patterns found for even or odd numbers of dart.

Example:

These students are considering the feasible scores for 5 darts using the target shown. One score might be 14 when the dart hits are $2 + 2 + 2 + 4 + 4$.

Logical Thinking

Extensions:

1. Have students use the dartboard shown below to answer the following questions, assuming that 6 darts are thrown unless otherwise noted: (1) Which of these scores are possible: 4, 19, 28, 58, 29, and 35? (2) What is the range of possible scores (the least and greatest)? (3) List all possible scores and demonstrate their correctness by showing the related addition (such as $30 = 3 + 3 + 5 + 5 + 7 + 7$)? (4) What kind of numbers were all of the achievable scores for 6 darts? Why? (5) If 5 darts instead of 6 were thrown, what scores would be possible and why?

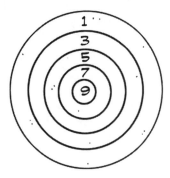

2. Have the students design their own targets and specify the number of darts to be thrown. Allow them to try out the proposed dartboards on one another, but also have discussions about the answers to such questions as those asked in Extension 1.

Dartboards

Chapter 103

Angelica's Bean Logic

Grades 4–8

☒ Total group activity
☒ Cooperative activity
☒ Independent activity
☒ Concrete/manipulative activity
☒ Visual/pictorial activity
☒ Abstract procedure

Why Do It:

Students will enhance their logical-thinking skills as they look for solution patterns.

You Will Need:

About 30 beans per player (or group) are needed, plus pencils and paper.

How To Do It:

Read the "Angelica's Bean Logic" problem to the group, provide beans, and allow the class to work together or individually to seek solutions. When a solution is found, it should be displayed (perhaps with an overhead projector) and recorded. The solver or solvers should share their solution procedures.

Angelica's grandfather always liked to tell her stories or have her try to solve puzzles. One day he said, "If you will count out 24 beans for me, I will show you some interesting tricks. I'm going to arrange the beans around a square with 3 in each group [see below]. That way you have 9 beans along each side of the square. Now what you must do is take 4 beans away and rearrange the rest so that you will still have 9 beans along each side of the square. Let me see if you can do it!"

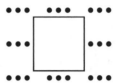

Example:

The students below have worked out one solution for the "Angelica's Bean Logic" problem. Ask participants to find two additional solutions. (*Note:* Further solutions are given later.)

I REMOVED 1 BEAN OUT OF THE CENTER GROUP ALONG EACH SIDE OF THE SQUARE. THEN I TOOK 1 MORE FROM EACH CENTER GROUP AND PLACED 2 OF THOSE IN THE UPPER LEFT CORNER GROUP AND 2 IN THE LOWER RIGHT CORNER GROUP. THAT DID IT!

IT SEEMS TO WORK, BUT HOW DID YOU FIGURE OUT THE SOLUTION?

Extensions:

Have students attempt the problems that follow.

1. Place the 24 beans around the square in groups of 3, as was done originally. This time, however, add 4 beans and still show 9 beans along each edge. See the original configuration below.

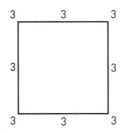

2. Try bean-square arrangements with 12 or 15 beans along each edge and see what patterns are possible.

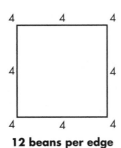 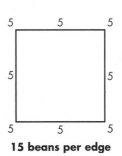

12 beans per edge   **15 beans per edge**

Solutions:

1. *Angelica's Bean Logic problem* (show 9 beans along each edge after removing 4):

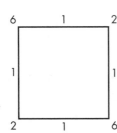 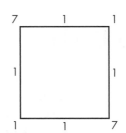

2. *Extension 1* (add 4 beans and still show 9 on each edge):

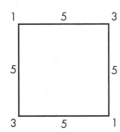 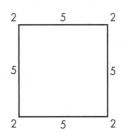

Chapter 104

Line It Out

Grades 4–8
- ☒ Total group activity
- ☒ Cooperative activity
- ☒ Independent activity
- ☒ Concrete/manipulative activity
- ☒ Visual/pictorial activity
- ☒ Abstract procedure

Why Do It:

Students will use their problem-solving and logical reasoning skills to investigate the intersection points of lines in a plane.

You Will Need:

Students will require "Line It Out" record-keeping sheets (included as a reproducible page), and pencils with erasers. Toothpicks or straws may also be used as optional aids.

How To Do It:

Line It Out can be an individual or cooperative effort. Help students solve the first two or three of the problems on the "Line It Out" worksheet provided. Some, such as Problem A-0, which calls for 2 lines with no intersections, can readily be solved with parallel lines. Others, such as Problem C-1, which asks for 4 lines with only one intersection, are sometimes more challenging (see Examples). Once students understand the activity's parameters, challenge them to find solutions for as many of the remaining problems on the "Line It Out" worksheet as they can. (*Note:* Remind them that lines go on forever and do not end within the boxes shown; an intersection outside of the box, therefore, must be considered as part of an answer.) Suggested solutions to the worksheet are provided at the end of the activity.

HEY! I FOUND A WAY TO CROSS 4 LINES AND GET 6 INTERSECTIONS.

Examples:

Shown below are solutions for Problems A-0, C-1, and D-5 on the "Line It Out" worksheet.

Show 2 LINES with
NO INTERSECTIONS

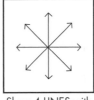

Show 4 LINES with
1 INTERSECTION

Show 5 LINES with
5 INTERSECTIONS

Extensions:

1. Give students this challenge question: With 5 lines (as in Problem Set D), is it possible to have more than 10 intersections?

2. Have able students create their own Problem Sets E through I for 6 through 10 lines.

3. Another challenge question that students might attempt is as follows: What solution or solutions might arise if these same questions on the "Line It Out" worksheet were considered in a 3-dimensional setting?

Line It Out

Line It Out

A. TWO LINES

INTERSECTION POINTS: 0—NONE   1—ONE

B. THREE LINES

INTERSECTION POINTS: 0—NONE   1—ONE   2—TWO   3—THREE

C. FOUR LINES

INTERSECTION POINTS: 0—NONE   1—ONE   2—TWO   3—THREE

4—FOUR   5—FIVE   6—SIX   7—SEVEN

D. FIVE LINES

INTERSECTION POINTS: 0—NONE   1—ONE   2—TWO   3—THREE   4—FOUR

5—FIVE   6—SIX   7—SEVEN   8—EIGHT   9—NINE   10—TEN

Logical Thinking

Some Solutions

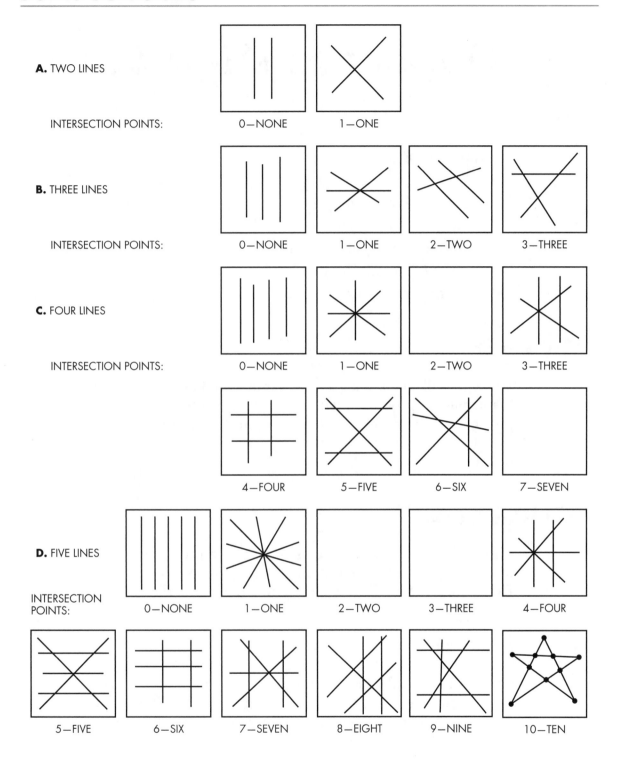

A. TWO LINES

INTERSECTION POINTS: 0—NONE   1—ONE

B. THREE LINES

INTERSECTION POINTS: 0—NONE   1—ONE   2—TWO   3—THREE

C. FOUR LINES

INTERSECTION POINTS: 0—NONE   1—ONE   2—TWO   3—THREE

4—FOUR   5—FIVE   6—SIX   7—SEVEN

D. FIVE LINES

INTERSECTION POINTS: 0—NONE   1—ONE   2—TWO   3—THREE   4—FOUR

5—FIVE   6—SIX   7—SEVEN   8—EIGHT   9—NINE   10—TEN

Chapter 105

Duplicate Digit Logic

Grades 4–8

☒ Total group activity
☒ Cooperative activity
☒ Independent activity
☐ Concrete/manipulative activity
☐ Visual/pictorial activity
☒ Abstract procedure

Why Do It:

Students will manipulate and compute duplicate digits in a mathematical problem involving adding, subtracting, multiplying, or dividing. They will use logical strategies in an effort to determine whether multiple, single, or no solutions are possible.

You Will Need:

This activity requires problems that can be presented on an overhead projector, on the chalkboard, or as photocopies. Each student should have paper and a pencil.

How To Do It:

1. Present the students with the following problem. Tell them that A, B, and C are different digits. Then challenge them to solve it in every possible way (one of multiple solutions is shown).

$$
\begin{array}{rr}
AAA & 111 \\
+BBB & +222 \\
\hline
CCC & 333 \\
\end{array}
$$

2. After the students have shown and discussed all possible solutions for the initial problem, present them with another that will stretch their logical-thinking abilities a bit further. Begin by asking which of the suggested solutions for DDD—111, 222, 333, 444, 555, and 666—are not possible, and why.

$$
\begin{array}{r}
AAA \\
BBB \\
+CCC \\
\hline
DDD
\end{array}
$$

3. Have the students also determine and list all of the possible solutions. Then have them consider whether the problem below can be solved.

Extensions:

Have students consider the following problems.

1. This problem involves subtracting duplicate digits. Determine all possible solutions (one example is provided).

$$
\begin{array}{r}
AAA \\
-BBB \\
\hline
CCC
\end{array}
\qquad
\begin{array}{r}
999 \\
-111 \\
\hline
888
\end{array}
$$

2. This problem involves palindromic (reversible) products of duplicate digits. Two problems with palindromic outcomes are shown below. What others are possible?

$$
\begin{array}{r}
33 \\
\times 11 \\
\hline
33 \\
33 \\
\hline
363
\end{array}
\qquad
\begin{array}{r}
222 \\
\times 111 \\
\hline
222 \\
222 \\
222 \\
\hline
24642
\end{array}
$$

3. Which division problems using duplicate digits have answers (quotients) with no remainders? Two examples are shown below. Are there other possibilities? List and discuss them.

$$
\begin{array}{r}
4 \\
222\overline{)888}
\end{array}
\qquad
\begin{array}{r}
3 \\
333\overline{)999}
\end{array}
$$

4. Activities for students may be extended beyond problems with duplicate digits to include such problems as the following (see Solutions below). Participants may also wish to create problems of their own. In this problem, each letter stands for a different digit, and students are to find numbers that work for the letters shown.

$$
\begin{array}{r}
DOG \\
+CAT \\
\hline
TOAD
\end{array}
$$

Solutions:

Extension 4 (there are other possible solutions):

302	403	706	807
+741	+621	+351	+261
1,043	1,024	1,057	1,068

Chapter 106

String Triangle Geometry

Grades 4–8

☒ Total group activity
☒ Cooperative activity
☒ Independent activity
☒ Concrete/manipulative activity
☒ Visual/pictorial activity
☒ Abstract procedure

Why Do It:

This hands-on activity allows students to construct geometric figures and to use concepts such as congruence, similarity, and parallelism to analyze the figures formed.

You Will Need:

Each group of students requires about 3 yards (or meters) of string; such measuring devices as rulers, yardsticks, or meter sticks; masking tape; pencils; paper; and scissors.

How To Do It:

In this activity, students will measure and manipulate string and tape as they practice their geometric problem-solving skills. It also provides students with an opportunity to apply their geometric identification and labeling skills.

1. Organize the class into groups of two or four students and give each group a piece of string approximately 3 yards long, about 10 inches of masking tape, a pencil, and scissors. Then instruct students to cut off 1/3 to

1/2 of their string and tape it to some flat surface (a tabletop, the chalkboard, or the floor, for example) in the form of any type of triangle. Each group should then label their triangle's vertices A, B, and C by writing in pencil on the tape (see the illustration).

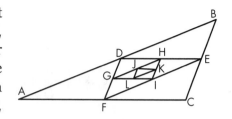

2. Next, have the students locate the midpoints of each edge of their triangles by any means that they desire (for example, by measuring or folding string). They should then connect those points with string and label the new vertices D, E, and F. When this is completed, ask, "What shapes do you see now?"

3. Students should then find the midpoints of the new triangle DEF and use string to connect them, labeling these points G, H, and I. Students repeat this process one more time, labeling the resulting triangle JKL. Then proceed to ask leading questions (answers are provided), such as

   a. How do triangles ADF and DBE compare?

   b. What can be said about triangles GHI and ABC?

   c. How many triangles the size of JKL are contained in triangle ABC?

   d. If triangle JKL were subdivided one more time into triangle MNO, how many triangles of that size would be included in triangle ABC?

   e. Can you determine a rule that will tell how many of the smaller triangles will result from each successive subdivision?

   f. What figures, other than triangles, are delineated by your strips?

4. When the students have spent sufficient time finding triangle-edge midpoints and connecting them to make subdivisions of the original triangle, suggest that they change their focus to extending the triangle outward by multiples. In order to accomplish this, students should begin with triangle PQR (see illustration below); they then extend PQ to S such that PQ = QS, and PR to T such that PR = RT.

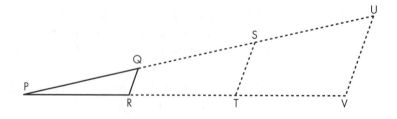

5. Then ask students the following questions:

   g. How many triangles the size of PQR would fit into quadrilateral RQST?

   h. How many times larger in area is triangle PST than triangle PQR?

6. Have students extend the sides of the triangle one multiple further such that SU = PQ and TV = PR, and proceed to ask more questions:

   i. How many triangles the size of PQR would fit into triangle PUV?

   j. Can you determine a rule that will tell us how many triangles the size of the original one (triangle PQR) there will be each time you expand outward by another multiple?

Solutions:

   a. ADF is congruent to DBE.

   b. ABC is similar to, and 16 times larger than, GHI.

   c. 64

   d. 256

   e. Multiply by 4 for each new subdivision.

   f. Figures include the parallelograms JKIL, DECF, and DHIG; the quadrilateral FDHI; and others.

   g. 3

   h. 4

   i. 9

   j. Square the number of original triangles plus the multiple extensions; thus, $1^2 = 1$ for triangle PQR; $2^2 = 4$ for triangle PST; and $3^2 = 9$ for triangle PUV. Then continue with $4^2, 5^2, 6^2, 7^2$, and so on.

Chapter 107

A Potpourri of Logical–Thinking Problems, Puzzles, and Activities

Why Do It:

These activities will provide students with a wide variety of logical-thinking and problem-solving experiences.

You Will Need:

A variety of easily obtained materials are cited within each of the activities that follow.

How To Do It:

See the individual activities below. Solutions are provided at the end.

Plan a Circuit Board

Grades K–8

☒ Total group activity
☒ Cooperative activity
☒ Independent activity
☒ Concrete/manipulative activity
☒ Visual/pictorial activity
☒ Abstract procedure

Many (perhaps most) of today's electronic circuits are printed rather than wired, making them perfect for 2-dimensional or plane geometry problems. On the circuit board illustrated here, the task is to connect terminals A and A, B and B, and C and C with printed electronic circuits

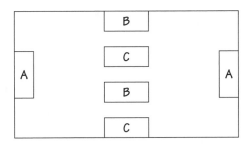

that do not touch. Should the electronic paths touch either each other or an incorrect terminal, they will short-circuit, and the device will malfunction. The goal is to draw circuit paths that connect the A terminals, the B terminals, and the C terminals without causing a short circuit.

22 Wheels and 7 Kids

Grades K–8

☒ Total group activity
☒ Cooperative activity
☒ Independent activity
☐ Concrete/manipulative activity
☒ Visual/pictorial activity
☒ Abstract procedure

Twenty-two wheels brought seven kids to school. They either walked, rode bicycles, or came in cars or trucks. Have students draw pictures to illustrate how the kids may have gotten to school and discuss their findings with the entire class or with a partner. (**Extensions:** Have students find the possibilities for 24 wheels and 8 kids, for 30 wheels and 9 kids, and so on.)

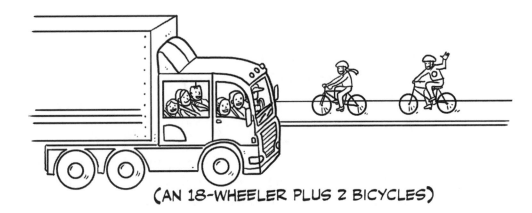

(AN 18-WHEELER PLUS 2 BICYCLES)

Logical-Thinking Problems, Puzzles, and Activities   **415**

Candy Box Logic

Grades 2–8
- ☒ Total group activity
- ☒ Cooperative activity
- ☒ Independent activity
- ☒ Concrete/manipulative activity
- ☒ Visual/pictorial activity
- ☒ Abstract procedure

The object of *Candy Box Logic* is to design candy boxes that will hold 36 pieces of candy and have no extra space. Students are to find all the possible ways for boxes that hold one, two, and three or more layers to contain 36 pieces. Have students draw pictures of their boxes or use blocks to show the different ways. (**Extensions:** Ask students to determine the possibilities, for example, for 12, 30, or 48 candies.)

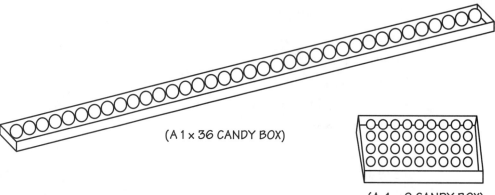

(A 1 × 36 CANDY BOX)

(A 4 × 9 CANDY BOX)

Brownie Cutting

Grades 2–8
- ☒ Total group activity
- ☒ Cooperative activity
- ☒ Independent activity
- ☒ Concrete/manipulative activity
- ☒ Visual/pictorial activity
- ☒ Abstract procedure

Give each student a brownie, and tell the class they can only eat their brownies once they have divided each one into 32 equal pieces using the lowest possible number of cuts. Have students first plan how they would make their cuts by drawing diagrams or thinking about the problem; then distribute the brownies and give students some time to work. Before they get to eat, have students both share how they divided the brownies and sketch the different methods on the chalkboard.

Making Sums with 0–9

Grades 2–8

☐ Total group activity
☒ Cooperative activity
☒ Independent activity
☐ Concrete/manipulative activity
☐ Visual/pictorial activity
☒ Abstract procedure

Each person will need a 3-digit addition sheet (as shown below) and matching 1-digit number cards for 0 through 9. Have each student remove one number card, perhaps with the numeral 3, and then use each of the remaining digits to construct a workable addition problem, finding and listing as many problems as they can. (**Extensions:** Students can remove different digits to find more workable problems. They can also create similar problems for subtraction, multiplication, or division.)

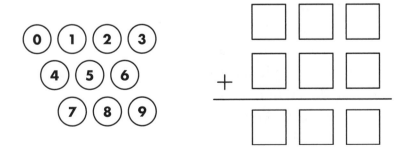

Logical-Thinking Problems, Puzzles, and Activities

Upside-Down Displays

Grades 2–8

☒ Total group activity

☒ Cooperative activity

☒ Independent activity

☐ Concrete/manipulative activity

☒ Visual/pictorial activity

☒ Abstract procedure

This activity involves using a hand-held calculator to display upside-down messages (see example below). Students first figure out what letter or letters each number (0–9) looks like when viewed upside down, and then create words or short messages from those letters. Next, students determine calculator computations that will yield the upside-down displays they planned, and try them out on other students.

$440 \times 7 = 3080$, but when read upside down we find a musical instrument

Coin Walk

Grades 2–8

☐ Total group activity

☒ Cooperative activity

☒ Independent activity

☒ Concrete/manipulative activity

☒ Visual/pictorial activity

☒ Abstract procedure

Taking a random Coin Walk requires 1 coin, a piece of graph paper for each student, and different-colored pencils or crayons. Begin at the lower-left corner of the graph paper and, for each toss of the coin, mark 1 unit to the right for a "head" or 1 unit up for a "tail." Have students predict where their random coin walk graphs will end. Continue the coin tosses and record the outcomes until the Coin Walk trail reaches an edge of the graph paper. Repeat the experiment two or three times using pencils of different colors. Ask students what logical statement might be made about the coin tosses.

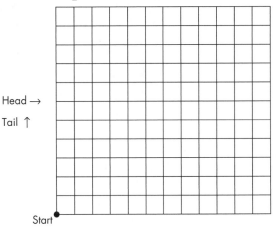

Dice Plotting

Grades 4–8

- ☐ Total group activity
- ☒ Cooperative activity
- ☒ Independent activity
- ☒ Concrete/manipulative activity
- ☒ Visual/pictorial activity
- ☒ Abstract procedure

Logical thinking and chance events both play roles in *Dice Plotting*. Place students in groups of two; each group will need a pair of red dice, a pair of green dice, a coordinate graph (as shown here), and pencils. The first student rolls 4 dice, 2 red and 2 green. He or she chooses 1 red and 1 green die, and marks the point (1,1) on the graph with an X. The second student then takes a turn and marks an O on the graph for his or her selected

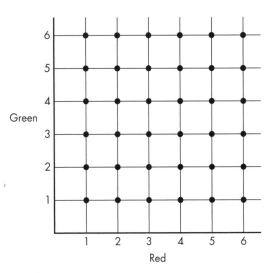

dice location. Once a point on the graph is marked, it belongs to that student. The winner is the student to get 4 marks in a horizontal, vertical, or diagonal row.

Logical-Thinking Problems, Puzzles, and Activities

Coin Divide

Grades 4–8

- [ ] Total group activity
- [x] Cooperative activity
- [x] Independent activity
- [x] Concrete/manipulative activity
- [x] Visual/pictorial activity
- [x] Abstract procedure

Place 18 coins (pennies are easiest) on grid paper as shown here. Challenge students to mark "fences" along the grid lines so that each fenced-in space has the same area and contains 3 coins.

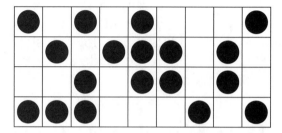

Animal Pens

Grades 4–8

- [ ] Total group activity
- [x] Cooperative activity
- [x] Independent activity
- [x] Concrete/manipulative activity
- [x] Visual/pictorial activity
- [x] Abstract procedure

In this problem scenario, a farmer has sheep in 3 large pens (A, B, and C). He needs to separate them in such a way that each animal will be in a pen of its own, but has only 3 lengths of portable fencing that he can use inside each of the large pens. Using toothpicks, students are required to form just 3 straight portable fence sections inside each of the large pens to separate the sheep so that each is in an individual pen.

A

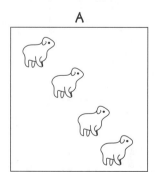

B

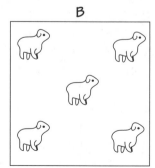

C

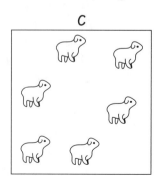

The farmer also had another strange pen situation. He told a friend that he had 15 pigs in 4 square pens, such that each pen contained an odd number of pigs. The friend said that was impossible, but then went to look and found it to be true. Have students determine how the farmer penned his pigs.

12 Days of Christmas

Grades 4–8

☒ Total group activity
☒ Cooperative activity
☒ Independent activity
☐ Concrete/manipulative activity
☒ Visual/pictorial activity
☒ Abstract procedure

According to the popular Christmas song, the following gifts were received successively during the 12 days of Christmas:

1st day	Partridge in a Pear Tree
2nd day	Turtle Doves
3rd day	French Hens
4th day	Calling Birds
5th day	Golden Rings
6th day	Geese a Laying
7th day	Swans a Swimming
8th day	Maids a Milking
9th day	Ladies Dancing
10th day	Lords a Leaping
11th day	Pipers Piping
12th day	Drummers Drumming

Have students determine:

- What was the total number of gifts?
- How many of the gifts were birds? (Optional: Answer as a fraction or percent.)
- How many gifts included people? (Optional: Answer as a fraction or percent.)
- What proportion of the gifts was jewelry?

Rubber Sheet Geometry

Grades 6–8

☐ Total group activity
☒ Cooperative activity
☒ Independent activity
☒ Concrete/manipulative activity
☒ Visual/pictorial activity
☒ Abstract procedure

Topology is a type of geometry in which the points, lines, and angles are permitted a great deal of motion. Figures in topology can shrink, stretch, bend, or be distorted. Because of this, topology has been nicknamed "Rubber Sheet Geometry." Students will be using *Rubber Sheet Geometry* to investigate maps and mapping situations.

This activity requires several pieces of thin, translucent rubber about 6 by 6 inches (this can be purchased from a drug store or made from cut-up rubber gloves); markers that will write on the rubber; thumb tacks; cardboard; a globe; several types of map projections; and a number of figures or words for tracing. Students begin by placing the rubber over a word and tracing it. Then they pull and stretch the rubber, observing what happens to the word. Although the word's length and width can be altered, and straight lines can be curved, the identifying portions (like the word COLA here), though distorted, remain in their constant relative positions (in the middle).

Now students should try a similar globe-and-map activity. They are to place a rubber sheet on a world globe (for example, North America), and trace a portion of it, including lines of longitude and latitude, on the rubber. Then they place the rubber sheet on a piece of cardboard and stretch it until the longitude lines are parallel to each other and perpendicular to the lines of latitude, securing the rubber with thumb tacks. The image created is a commonly used projection that is most often termed a Mercator map. A Mercator projection is a map projection, where a three-dimensional map is put on a two-dimensional surface. (**Extension:** Students might research and construct other types of projections, such as azimuthal, conic, cylindrical, or homolographic projections. For more on map projections, students can consult http://en.wikipedia.org/wiki/Map_projection.)

How Long Is a Groove?

Grades 6–8
- ☒ Total group activity
- ☒ Cooperative activity
- ☒ Independent activity
- ☒ Concrete/manipulative activity
- ☒ Visual/pictorial activity
- ☒ Abstract procedure

Obtain a large bolt and inspect its threading or grooves. Ask students to guess what the total length of that groove might be and how they could find out. For a more challenging activity, ask, "What is the diameter of the bolt? Is the diameter in the groove the same? Could you make a calculation from these figures? Could you use string to find the circumference for 1 rotation? How many rotations does the groove make?" Have students calculate and compare findings with a partner. They are to use a long piece of string, wrapping it through the entire groove, marking it, unwrapping it, measuring it, and comparing it to their calculations to see how close their measurements came. (**Extensions:** Students can use similar methods to determine the length of a groove on a long-playing vinyl record, or to find the length of the tape in an old cassette tape or VHS tape.)

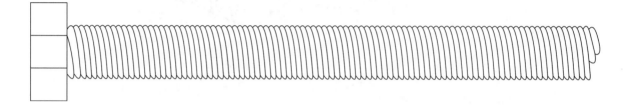

Solutions to Selected Potpourri Activities:

Plan a Circuit Board

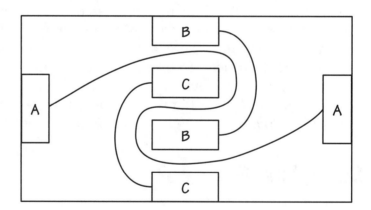

22 Wheels and 7 Kids

Any workable solution is acceptable. The following are possibilities:

- 4 kids came in separate cars + 3 rode bicycles = (4 × 4) + (3 × 2) = 22
- 5 kids came in cars + 1 rode a bicycle + 1 walked = (5 × 4) + (1 × 2) + 0 = 22
- 5 kids rode in my Dad's 18-wheeler + 2 rode bikes = 18 + (2 × 2) = 22

Logical Thinking

Candy Box Logic

The 1-layer boxes for 36 candies will range from 1 by 36, to 2 by 18, to 3 by 12, to 4 by 9, to 6 by 6 arrangements. (*Note:* The participants are dealing with all of the multiplication facts for 36.)

Brownie Cutting

Twelve and thirteen cuts are quite interesting, ten cuts is the usual solution, but the most efficient solution is 7 cuts.

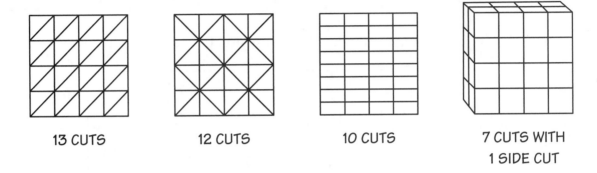

13 CUTS   12 CUTS   10 CUTS   7 CUTS WITH
                                                    1 SIDE CUT

Upside-Down Displays

A few additional calculator computations that yield upside-down messages are:

- 52,043 ÷ 71 and get a snake like fish — EEL
- 159 × 357 − 19,025 and get a beautiful young lady — BELLE
- 161,616 ÷ 4 and get what Santa might say — h0h0h0
- 2,101 × 18 and get the name of a good book — BIBLE
- $73^2 + 9$ and get a honey of an answer — BEES

Coin Divide

The following is one possible solution.

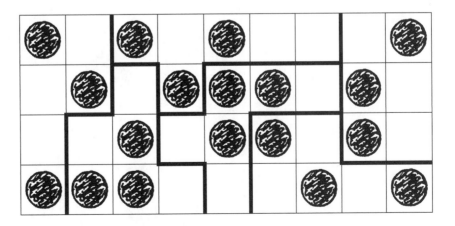

Logical-Thinking Problems, Puzzles, and Activities

Animal Pens

The following are possibilities for separating the sheep in the pens A, B, and C.

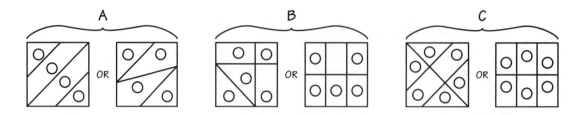

The pig pens might have been situated as shown below:

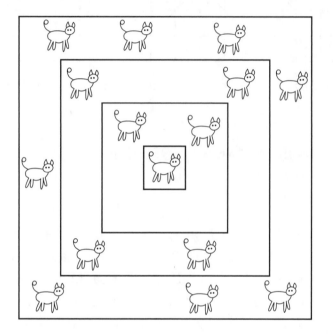

Selected Bibliography

California State Board of Education. (2006). *Mathematics Framework for California Public Schools: Kindergarten Through Grade Twelve*. Sacramento: California Department of Education.

National Council of Teachers of Mathematics. (1991). *Algebra for Everyone* [video recording]. Reston, VA: Mathematics Education Trust.

National Council of Teachers of Mathematics. (2006). *Principles and Standards for School Mathematics*. Retrieved January 10, 2007, from www.nctm.org/standards/overview.htm.

National Research Council, Donovan, S. M., and Bransford, J. D. (eds.). (2005). *How Students Learn History, Mathematics and Science in the Classroom*. Washington, DC: The National Academies Press.

Overholt, J. L. (1978). *Dr. Jim's Elementary Math Prescriptions*. Glenview, IL: Goodyear/Scott, Foresman.

Overholt, J. L., Aaberg, N. H., and Lindsey, J. F. (2008). *Math Stories for Problem Solving Success* (2nd ed.). San Francisco: Jossey-Bass.

Overholt, J. L., White-Holtz, J., and Dickson, S. S. (1999). *Big Math Activities for Young Children*. Albany, NY: Delmar/International Thomson.

Utah State University. (2007). *National Library of Virtual Manipulatives* [electronic version]. Retrieved January 5, 2009, from www.nlvm.usu.edu/en/nav/index.html.

Index

Index